MATHEMATICAL RESEARCH | MATHEMATISCHE FORSCHUNG

Structural Induction on Partial Algebras

H. Reichel

Band 18

Akademie-Verlag · Berlin

In this series original contributions of mathematical research in all fields are contained, such as

- research monographs
- collections of papers to a single topic
- reports on congresses of exceptional interest for mathematical research.

This series is aimed at promoting quick information and communication between mathematicians of the various special branches.

In diese Reihe werden Originalbeiträge zu allen Gebieten der mathematischen Forschung aufgenommen wie

- Forschungsmonographien
- Sammlungen von Arbeiten zu einem speziellen Thema
- Berichte von Tagungen, die für die mathematische Forschung besonders aktuell sind.

Die Reihe soll die schnelle Information und gute Kommunikation zwischen den Mathematikern der verschiedenen Fachgebiete fördern.

Manuscripts
in English and German comprising at least 100 pages and not more than 500 pages can be admitted to this series. With respect to a quick publication the manuscripts are reproduced photomechanically. Authors who are interested in this series please turn directly to the 'Akademie-Verlag . Here you will get more detailed information about the form of the manuscripts and the modalities of publication.

Manuskripte
in englischer und deutscher Sprache, die mindestens 100 Seiten und nicht mehr als 500 Seiten umfassen, können in diese Reihe aufgenommen werden. Im Interesse einer schnellen Publikation werden die Manuskripte auf fotomechanischem Weg reproduziert. Autoren, die an der Veröffentlichung entsprechender Arbeiten in dieser Reihe interessiert sind, wenden sich bitte direkt an den Akademie-Verlag. Sie erhalten dort genauere Informationen über die Gestaltung der Manuskripte und die Modalitäten der Veröffentlichung.

H. Reichel

Structural Induction on Partial Algebras

Mathematical Research · Mathematische Forschung

Wissenschaftliche Beiträge
herausgegeben von der
Akademie der Wissenschaften der DDR
Institut für Mathematik

Band 18

Structural Induction on Partial Algebras

by
H. Reichel

Structural Induction on Partial Algebras

Introduction to Theory and Application of Partial Algebras – Part II

by

Horst Reichel

Akademie-Verlag · Berlin 1984

Autor:

Prof. Dr. Horst Reichel
Technische Hochschule "Otto von Guericke"
3010 Magdeburg
Boleslav-Bierut-Platz 5

Die Titel dieser Schriftenreihe werden vom Originalmanuskript der Autoren reproduziert.

ISSN 0138 - 3019

Erschienen im Akademie-Verlag, DDR-1086 Berlin, Leipziger Str. 3-4

Lizenznummer: 202 • 100/405/84
Printed in the German Democratic Republic
Gesamtherstellung: VEB Kongreß- und Werbedruck, 9273 Oberlungwitz
Lektor: Dr. Reinhard Höppner
Umschlag: Karl Salzbrunn
LSV 1025
Bestellnummer: 763 363 0 (2182/18)
02600

PREFACE

Our purpose in writing this book is an increasing use of partial algebras in computer science. We hope to achieve this aim by the development of a special kind of partiality, the so-called equational partiality, as a sound mathematical basis for algebraic specifications using partial operations and as one of several possible generalizations of the Universal Algebra to partial algebras.

The book presents both the developed theory of equational partiality and its applications in algebraic specifications. The most important concepts with respect to applications are the notions of a canon and of the recursive dependence within canons. The usefulness of partial algebras is proved by the fact that structural induction alone yields on partial algebras a complete formal calculus of abstract programs on abstract parameterized data types.

The presentation of the complete theory of partial algebras is out of the range of this book. Therefore the author cooperates with Peter Burmeister in writing an

"Introduction to Theory and Application of Partial Algebras"

consisting of two parts:

Part I : "A Model Theoretic Approach to Partial Algebras"
by Peter Burmeister , and

Part II: "Structural Induction on Partial Algebras"
by Horst Reichel.

The content of the present book is the result of a long joint work with Klaus Benecke, Ulrich L. Hupbach, and Heinz Kaphengst, and I am very thankful to these colleagues for that collaboration.

I would also like to thank Rod Burstall and Don Sannella for helpful discussions and suggestions during my Visiting Fellowship at Edinburgh University made possible by a research grant of SERC.

Thanks are due to Wolfgang Wechler for careful proof-reading and useful comments concerning the manuscript.

Finally, I want to thank the Akademie-Verlag for friendly cooperation and quick publication.

Magdeburg, October 1983 H.Reichel

C O N T E N T S

INTRODUCTION

The main purpose of the book is the initiation of an increasing use of partial algebras in computer science. There are many reasons for the inadequate state of application of partial algebras. Probably, the bad reputation of the theory of partial algebras itself causes that inadequate state.

The theory of partial algebras was initiated for mathematical reasons more or less independently by J. Schmidt and his group and by J. Slominski. In the years from 1965 to 1973 many important results were contributed by P. Burmeister, R. Kerkhoff, H. Höft and V.S. Poythress. But this first and purely mathematical impulse did not initiate a continuously increasing development of the theory of partial algebras. The first results showed that a straightforward generalization of the theory of total algebras was not to be supposed. Many different generalizations of the concept of homomorphism, of substructure, and of validity of term equations were introduced, and it was not quite clear which of them would be the right ones. At this time there were not sufficiently many examples of partial algebras by means of which one could decide these questions.

Meanwhile the extensively developed theory of small categories and to a greater extend the application of universal algebra in computer science act as motives for the development of the theory of partial algebras. In the present Part II of the 'Introduction to Theory and Application of Partial Algebras' we present a special theory of partial algebras mainly oriented towards satisfying the requirements of applications of universal algebra in computer science and of the theory of small categories. The variety of possible generalizations of the theory of total algebras will be thoroughly discussed in Part I 'A Model Theoretic Approach to Partial Algebras' by P. Burmeister.

Next we try to explain the basic idea of the presented theory of 'EQUATIONAL PARTIALITY'. One reason for the frequent application of algebra in computer science is the use of structural induction on free algebraic structures. Very often one has to define a mapping $f: M \longrightarrow P$ in a finite manner, where M is an infinite but finitely representable set. A typical example of this problem is the definition of the semantics of a programming language. In this case the set M consists of all syntactically correct programs and P is a set of state transformations of an abstract interpreting machine. The finite representation of M is given by a phrase structure grammar.

An application of structural induction to define a function $f: M \longrightarrow P$ may be sketched in the following way: at first one has to find a finite algebraic theory T in such a way that M becomes the carrier set of an initial T-algebra $\mathbb{M}$, where a T-algebra $\mathbb{A}$ is said to be initial if for every T-algebra $\mathbb{B}$ exactly one homomorphism $f: \mathbb{A} \longrightarrow \mathbb{B}$ exists. In the second step of the inductive definition of $f: M \longrightarrow P$ we have to define on the set P any T-algebra $\mathbb{P}$ so that the desired function is given by the uniquely determined homomorphism $f: \mathbb{M} \longrightarrow \mathbb{P}$.

Intuitively a finite representation of an infinite set M by means of a finite algebraic theory T can be understood as follows: A finite algebraic theory T is given by a finite set Ω of operations and by a finite set $\mathcal{O}\!\!\!\mathcal{L}$ of axioms describing which combinations of the fundamental operations produce the same value. Among the fundamental operations the nullary operations have the function of atomic generating elements of M, so that any element of M can be represented as the value of a finite combination of non-nullary operations applied to atomic generating elements. If any two such combinations produce the same value then this must be a consequence of the axioms. In the simplest case the theory T can be chosen in such a way that the set of axioms is empty. If we take the simplest non-trivial case, given by one atomic generating element, by one unary generating operation, and by an empty set of axioms, we get exactly Peano's axiomatization of natural numbers. Therefore, mathematical induction on natural numbers is a special case of structural induction.

In the case of formal languages any production rule of a context - free grammar, containing no non-terminal in its right hand side, can be identified with a constant operation, or an atomic generating element, and any other production rule can be interpreted as a generating operation. The existence of different non-terminals corresponds in algebraic terms to different kinds of carrier sets, e.g. we have to use many-sorted or heterogeneous algebras. Combinations of generating operations correspond to derivation trees. If the grammar is unambiguous the sets of derivable strings form a free heterogeneous algebra, so that the semantic meaning can be defined by structural induction. The definition of the algebra $\mathbb{B}$, which is the range of the semantic interpretation itself, corresponds in the context of programming languages to a description of an interpreter by giving a semantic interpretation to every production rule. More detail in this direction can be found in the paper of Knuth (1968) or in a paper of Letitshevski (1968) and in the paper 'Initial Algebraic Semantics' of the ADJ group (1975). However, the syntactic structure of existing programming languages cannot be completely described in an adequate way by context-free grammars. In general, most production rules are restricted by so-called context conditions. Algebraically we have to interpret a production rule with a context condition as a partial operation with a necessary and sufficient domain condition.

The same phenomenon can be observed in small categories. Small categories have exactly one partial operation, namely the partial composition of morphisms. And again, the partial binary operation is equipped with a necessary and sufficient domain condition. The product $x \circ y$ exists if and only if the target object of x is equal to the source object of y.

According to these observations in the theory of equationally partial algebras we equip each operator with a necessary and sufficient domain condition, which is given by a finite set of term equations. A total operator is then an operator equipped with a trivial domain condition, for instance with the empty set of term equations. We shall develop the theory of equationally partial structures in such a way that the notions and results of total algebras will be reproduced if every operator has a trivial domain condition. We shall prove that a certain syntactic hierarchy condition for the domain conditions signatures (or of heterogeneous operator schemes) is necessary and sufficient for a nice and simple theory of equational partiality. With this approach we can describe the inductive power of context-restricted production rules of real programming languages.

The application of structural induction on partial algebras will be demonstrated in Chapter one by specifications of several abstract data types and of parameterized abstract data types containing partial operations. We present these applications of structural induction on partial algebras without any preceding theoretical explanations in order to show that those specifications are not significantly harder then specifications using total operations only.

In the following three chapters we develop thoroughly the algebraic foundations of structural induction, of parameterized abstract data types with partial operations, and of functional enrichments of parameterized abstract data types.

In Chapter two we develop the algebraic theory of equational partiality. The basic notions are hierarchical equationally partial signatures (hep-signatures), hep-signatures with elementary implications as axioms, and finally hep-quasi-varieties as classes of models of hep-signatures with elementary implications. It will be shown that due to the hierarchy condition of the system of domain conditions the weakest concept of homomorphism between partial algebras is sufficient for the development of a sound theory of equationally partial algebras that for instance completely reflects the structural behaviour of small categories. In this respect Chapter two mainly contributes to the Universal Algebra of partial algebras.

Chapter three develops the algebraic foundations of structural induction on partial algebras. It is mainly based on the existence of partial algebras that are freely generated by sets of generators and sets of defining relations on the generating elements. This concept, well-known in the theory of groups, is the algebraic key for a powerful and flexible approach to structural induction which works as well for total algebras as for partial algebras. This concept can also be used for constructions of partial quotient algebras. The theorem of homomorphisms in Section 3.4 represents the most significant difference in the structural behaviour of total algebras and of equationally partial algebras.

In Chapter four we deal with classes of equationally partial algebras that are exlusively motivated by semantic problems of parameterized abstract data types and by functional enrichments of parameterized abstract data types. These model classes are defined by so-called CANON's , or by initially restricting algebraic theories. On the basis of this concept we define computable enrichments and definable enrichments of canons and prove some necessary conditions for definable and for computable enrichments of canons. In the last section we show how canons and canon morphisms can be used for the definition of the semantics of constructions that compose complex specifications from simpler ones. Finally we consider constructions like instantiations of parameterized abstract data types that preserve computable enrichments of canons.

CHAPTER 1

SPECIFICATIONS OF ABSTRACT DATA TYPES WITH PARTIAL OPERATIONS

1.1 Abstract Data Types - ADT's

In connection with investigations of modularity concepts of large software projects the understanding of the data type concept was essentially modified. Whereas in programming languages like ALGOL 60, PL/1, and PASCAL a data type is considered to be a set, at least since the early seventies it seems to be generally accepted that data types are not just sets, but sets equipped with operations. For instance, if we deal with strings the following operations can be considered to be relevant:

$$(*)\quad \begin{cases} \text{equal(String,String)} \longrightarrow \text{Bool} \\ \text{append(String,Character)} \longrightarrow \text{String} \\ \text{interlock(String,String)} \longrightarrow \text{String} \\ \text{contained(Character,String)} \longrightarrow \text{Bool} \\ \text{length(String)} \longrightarrow \text{Nat} \\ \text{empty} \longrightarrow \text{String} \end{cases}$$

Here we see that four different kinds of objects occur which are characterized by the type names 'Bool', 'String', 'Nat', and 'Characters', respectively. This situation can algebraically be represented by a 4-sorted algebra A consisting of four carrier sets, namely $A_{Bool} = \{T,F\}$, $A_{Nat} = \mathbb{N}$ (set of natural numbers), $A_{Characters} = \{a,b,c,\dots,y,z\}$, and A_{String} = set of all strings over $A_{Characters}$, and which consists of six fundamental operations:

$$\text{equal}^A\colon A_{String} \times A_{String} \longrightarrow A_{Bool}$$

$$\text{append}^A\colon A_{String} \times A_{Characters} \longrightarrow A_{String}$$

$$\text{interlock}^A\colon A_{String} \times A_{String} \longrightarrow A_{String}$$

$$\text{contained}^A\colon A_{Characters} \times A_{String} \longrightarrow A_{Bool}$$

$$\text{length}^A\colon A_{String} \longrightarrow A_{Nat}$$

$$\text{empty}^A\colon \{\emptyset\} \longrightarrow A_{String}$$

Depending on the choice of the actual alphabet, i.e. depending on the interpretation of the type name 'Characters', we would get different concrete data types.

The example demonstrates that a concrete data type can be identified with a many-sorted algebra. But this understanding is not sufficient to put into practice the suggestion of Hoare (1972): "In the development of programs by stepwise refinement the programmer is encouraged to postpone the decision on representation of his data until after he has designed his algorithm and has expressed it as an abstract program operating on abstract data." The problem consists in finding a precise definition of an abstract data type and of an abstract program operating on an abstract data type.

The fundamental idea in abstract data type specifications is the mathematical presentation of an abstract data type by a class of behaviourally equivalent many-sorted algebras. In the literature on abstract data types there are several different definitions of behavioural equivalence. Here we use the mathematically most simple version of equivalence which considers many-sorted algebras as behaviourally equivalent if they are isomorphic. In Chapter four we will make precise the notion of an abstract program operating on an abstract data type.

However, there are many very different interpretations of the preceding scheme (*) which have nothing in common with the intended model of strings over an alphabet. Although schemes of the form (*) are used in the programming language ADA as specification part of a package it has no semantics. In ADA the semantics of a package is given by the implementation, or by the body of the package. This approach does not represent the behaviour of an abstract data type by the description of the properties of the available manipulation functions, but by giving a representant of the behavioural class defined in terms of other ADA packages and of the basic types and basic functions of ADA. A disadvantage of this approach can be illustrated by the following example. If the natural numbers would be represented by the special model of Roman numbers, i.e. I,II,III,IV,V,VI,... , everybody sees the difficulties in discovering the distributivity law of addition and multiplication of natural numbers. Therefore we prefer axiomatic tools in descriptions of the external behaviour of abstract data types.

Since our aim is the demonstration of the usefulness and feasibility of structural induction on partial algebras we will not explain the fundamentals of algebraic specifications of abstract data types. Readers who are not familiar with the basic ideas should consult the paper 'Introduction to Data Algebra' by S.N. Zilles in LNCS 86 (1980) or 'A Look at Algebraic Specifications' by the same author (see S.N. Zilles (1981)).

Coming back to the preceding example one can easily check that the intended interpretation satisfies the following properties expressed by the equations:

$$
(**)\quad
\begin{array}{l}
\text{length(empty)} = 0\\
\text{length(append(w,x))} = \text{length(w)} + 1\\
\text{interlock(w,empty)} = w\\
\text{interlock(w,append(v,x))} = \text{append(interlock(w,v),x)}
\end{array}
$$

where w,v are variables for strings and x is a variable for characters.

The next interpretation of the scheme (*) shows that the system (**) of equations is not strong enough to exclude all those interpretations which are not isomorphic to the intended interpretation. Let the interpretation B be given by

$B_{Bool} = A_{Bool} = \{T,F\}$, $B_{Nat} = A_{Nat} = \mathbb{N}$,

$B_{Characters} = A_{Characters}$,

B_{String} = set of all functions $f: B_{Characters} \longrightarrow \mathbb{N}$,

$empty^B = f_o$ with $f_o(x) = 0$ for all $x \in B_{Characters}$,

$append^B(f,x) = f'$ with $f'(y) = f(y)$ for $y \neq x$ and $f'(x) = f(x) + 1$ for $x,y \in B_{Characters}$,

$interlock^B(f,g) = h$ with $h(x) = f(x) + g(x)$, $x \in B_{Characters}$,

$$contained^B(x,f) = \begin{cases} T & \text{iff } f(x) \neq 0 \\ F & \text{iff } f(x) = 0 \end{cases} \quad \text{for } x \in B_{Characters} ,$$

$length^B(f) =$ $f(x)$ over all $x \in B_{Characters}$,

$$equal^B(f,g) = \begin{cases} T & \text{iff } f(x)=g(x) \text{ for all } x \in B_{Characters} \\ F & \text{iff } f(x) \neq g(x) \text{ for some } x \in B_{Characters} \end{cases} ,$$

The interpretation B represents multi-sets and does not represent strings. Evidently, A and B are not isomorphic since there is an equation which is satisfied by B but not by A , namely

$$\text{append(append(w,x),y)} = \text{append(append(w,y),x)}.$$

This example indicates that we need an additional tool for excluding undesirable interpretations.

This additional exclusion of interpretations will be done by the concept of 'initial satisfaction of axioms. Informally we can say that an interpretation C of (*) initially satisfies the set (**) of equations, if C satisfies only those equations which are consequences of (**) . Where an equation $t_1 = t_2$ is called a consequence of a set G of equations if every algebra D satisfying each equation in G also satisfies $t_1 = t_2$. Interpretation B shows that the equation

append(append(w,x),y) = append(append(w,y),x)

is not a consequence of (**), so that B does not initially satisfy the set of equations given by (**).

As next we will make the concept of initial satisfaction more precise. Readers wanting a more detailed explanation should look into the paper 'An Initial Algebra Approach to the Specification, Correctness, and Implementation of Abstract Data Types' of the ADJ group (see ADJ (1978)).

In terms of universal algebra one can say that an algebra C of scheme (*) initially satisfies (**), iff for every algebra D of scheme (*) satisfying (**) there is exactly one homomorphism $f: C \longrightarrow D$. Obviously any two algebras are isomorphic if they initially satisfy (**) . This shows that initial satisfaction fits for specifications of isomorphism classes or in other words for definitions of abstract data types.

But, it was shown by Bergstra and Meyer (1983) that there are very simple abstract data types that are not specifiable by finite sets of equations. They show that the set of all finite subsets of natural numbers with the cardinality function may serve as such a counter-example. A finite specification of that data type requires conditional equations of the form

$$t_1 = r_1, \dots, t_n = r_n \longrightarrow t = r .$$

As next we introduce the syntactic form for writing specifications of abstract data types by means of a simplified version of strings.

```
STRING  is definition
        sorts  Nat, String, Characters
        oprn   zero  ──→ Nat
               succ(Nat)  ──→ Nat
               a,b,...,y,z  ──→ Characters
               append(String,Characters)  ──→ String
               length(String)  ──→ Nat
        axioms  w,v:String, x:Characters
        length(empty) = zero
        length(append(w,x)) = succ(length(w))
        interlock(w,empty) = w
        interlock(w,append(v,x))=append(interlock(w,v),x)
        end STRING
```

The key-word <u>definition</u> indicates initial satisfaction for the following axioms. After the key-word <u>sorts</u> the new sort names follow which correspond to the specified data types. After the key-word <u>oprn</u> the sequence of fundamental operators follow, where in brackets, following the operator name, the number, the ordering and the required sorts of the arguments of that operator are declared by an n-tuple of sort names. The sort name after the arrow declares the result sort. This notation is intentionally similiar to function declarations in programming languages. This implies that in the preceding specification 'zero ⟶ Nat' denotes a constant of type 'Nat', i.e. an operator without any arguments that produces an object of type 'Nat', and that

append(String, Characters) ⟶ String

denotes an operator which needs two arguments, the first is of sort 'String' and the second of sort 'Characters', and the resulting value is again of sort 'String'.

a,b,...,y,z ⟶ Characters

declares a set of constant operators with the names a,b,...,y,z all of type 'Characters'.

After the key-word <u>axioms</u> the declaration of the variables follows that are used in the following axioms. In the preceding specification each axiom is an equation.

A verification of the specification STRING would require the proof of initiality of the interpretation which is suggested by the chosen names of sorts and of operators.

The next example illustrates the use of conditional equations by a specification of the set of natural numbers, of the set of all finite subsets of natural numbers equipped with the cardinality function.

```
CARDINALITY is definition
        sorts  Nat, Sets, Bool
        oprn   zero ⟶ Nat
               succ(Nat) ⟶ Nat
               true, false ⟶ Bool
               eq(Nat,Nat) ⟶ Bool
               ∅ ⟶ Sets
               Sets U {Nat} ⟶ Sets
               card(Sets) ⟶ Nat
               Nat ∈ Sets ⟶ Bool
        axioms  x,y: Nat,  m: Sets
        eq(x,x) = true
        eq(x,y) = eq(y,x)
        eq(zero,succ(x)) = false
        eq(x,y) = eq(succ(x),succ(y))
```

$(m \cup \{x\}) \cup \{x\} = m \cup \{x\}$

$(m \cup \{x\}) \cup \{y\} = (m \cup \{y\}) \cup \{x\}$

$(x \in \emptyset) = \text{false}$

$x \in (m \cup \{x\}) = \text{true}$

if $(y \in m) = \text{false}$ then $y \in (m \cup \{x\}) = eq(x,y)$

$card(\emptyset) = zero$

if $(x \in m) = \text{false}$ then $card(m \cup \{x\}) = succ(card(m))$

end CARDINALITY

In order to improve the readability of specifications we allow besides the functional notation of operator application as well the infix notation, as it is done in the case 'Nat $\in$ Sets $\longrightarrow$ Bool', as the mixed-fix notation, which is used for the operator of adding a natural number to a finite set ' Sets $\cup$ {Nat} $\longrightarrow$ Sets '.
A further step in this direction could be the abbreviation of an equation '(x rel y) = true' by 'x rel y' and of an equation '(x rel y) = false' by 'x rel y' , if 'rel' is a name of a binary infix operator producing truth values. Using this convention some of the preceding axioms could be written as follows:

$x \notin \emptyset$, $x \in (m \cup \{x\})$,

if $x \notin m$ then $y \in (m \cup \{x\}) = eq(x,y)$,

if $x \notin m$ then $card(m \cup \{x\}) = succ(card(m))$.

Assuming that the preceding examples sufficiently explain the way of writing specifications of abstract data types with total operations we turn now to specifications of data types with partial operations.

Without any algebraic foundation partial operations are extensively used in attribute grammars in the form of restricted production rules, where the restrictions of production rules may be expressed by equations on attributes. By the following examples we will demonstrate that the initial approach of the definition of syntax and semantics of context-free languages can be extended to languages generated by a grammar with restricted production rules. Also in this case syntax and semantics can be defined via homomorphisms from an initial partial algebra built up from derivation trees of the grammar.

To start with simpler examples we firstly specify the set of all those finite sequences of natural numbers that contain each natural number at most once. For an inductive construction of all those sequences the adding of a natural number n to a sequence r will be restricted to the case that n is not contained in r . This leads to a partial operation.

Since we will need natural numbers for several specifications we start with the specification NAT which will be extended to EXAMPLE1 afterwards.

```
NAT is definition
        sorts  Nat, Bool
        oprn   true, false ⟶ Bool
               zero ⟶ Nat
               succ(Nat) ⟶ Nat
               eq(Nat,Nat) ⟶ Bool
        axioms  x,y: Nat
        eq(x,x) = true
        eq(x,y) = eq(y,x)
        eq(zero,succ(x)) = false
        eq(succ(x),succ(y)) = eq(x,y)
end NAT

EXAMPLE1  is  NAT  with definition
        sorts  Rseq
        oprn   nil ⟶ Rseq
               Nat ∈ Rseq ⟶ Bool
               add(x:Nat,r:Rseq iff x ∉ r) ⟶ Rseq
        axioms  x,y: Nat,  r: Rseq
        x ∉ nil
        if x ∉ r then   y ∈ add(x,r) = eq(x,y)
end EXAMPLE1
```

In the specification EXAMPLE1 the specification NAT is used as prerequisite. In this way complex specifications can be written by stepwise enrichments of presupposed specifications.

Of course, this example could also be specified with a total adding operator. In this case we would have to add a new axiom that guarantees that the addition of a natural number n contained in the sequence r does not change the sequence r , i.e.

```
        if  x ∈ r  then  add(x,r) = r  .
```

Another way of making the adding operator to a total one would be the use of a so-called error value, as it was suggested by Goguen (1977). The resulting specification is then:

```
EXAMPLE2  is  NAT  with definition
        sorts  Rseq
        oprn   nil ⟶ Rseq
               error ⟶ Rseq
```

```
            add(Nat,Rseq) ——> Rseq
      axioms x,y: Nat, r: Rseq
      x ∉ nil
      x ∉ error
      if x ∉ r then y ∈ add(x,r) = eq(x,y)
      add(x,error) = error
      if x ∈ r then add(x,r) = error
end EXAMPLE2
```

It is evident that these three specifications do not specify one and the same abstract data type. It is also meaningless to say that only one of these specifications is the correct one. It depends on the context of application in which the abstract data type will be used. Only from this context one gets the information to decide whether or not a programmer should be forced to think about the applicability of the add-operator before using that operator, or if the programmer should provide exception handling which comes into account after an application of the add-operator, or if it is finally of no interest from outside if a programmertries the addition of a natural number to a sequence that is already contained in the sequence. In the last case the occurrence of such an addition would be detected by the abstract data type itself and according to the axiom '<u>if</u> x ∈ r <u>then</u> add(x,r) = r' the abstract data type would act correctly.

This discussion shows that any specification language should give the designer of a software project the possibility to decide in which way partiality is considered. This is one reason why we claim that a specification language should allow the specification of partial operations. Furthermore it is proved in the theoretical considerations of Section 3.4 that there is no way to describe the structural behaviour of partial algebras in an adequate way in terms of total algebras. Thus, there are practical and theoretical reasons to include partial operations in specification languages.

In the next example we specify trees whose edges are valuated with natural numbers in sucha way that each path from the root to any leaf induces a decreasing sequence of edge values. This specification is devided intotwo part. In the first part we do not specify any new type, but we enrich NAT by additional operations that are very useful for the following specification of 'decreasing edge-valued trees'.

```
DECREASING EDGE-VALUED TREES is NAT with definition
        oprn   max(Nat,Nat) ——> Nat
               Nat < Nat  ——>  Bool
```

```
axioms  x,y: Nat
x ≮ zero
zero < succ(x)
(x < y) = ( succ(x) < succ(y))
max(x,x) = x
max(x,y) = max(y,x)
if x < y  then  max(x,y) = y
with definition
sorts  Tree, Treeseq
oprn  nil ——→ Treeseq
      add(Treeseq,r:Tree,n:Nat iff t-attr(r) <n) —e→ Treeseq
      combine(Treeseq) ——→ Tree
      t-attr(Tree) ——→ Nat
      seq-attr(Treeseq) ——→ Nat
axioms  n: Nat, r: Tree, s,s1,s2: Treeseq
seq-attr(nil) = zero
t-attr(combine(s)) = seq-attr(s)
if s1=add(s2,r,n) then seq-attr(s1) = max(seq-attr(s2),n)
```

end DECREASING EDGE-VALUED TREES

This specification is divided into two parts. In the first part no new sort name is introduced, i.e. no new type is specified. Therefore the first part for itself is only a functional extension of NAT . This functional enrichment serves as preparation for the specification of the intended data type.

The formal expression 'combine(nil)' represents the unique tree consisting of exactly one node and of no one edge. 'combine(nil) will be pictured by a small circle. The objects of sort 'Treeseq' are special sequences of pairs constituted by a tree and by a natural number. As well for trees as for objects of sort 'Treeseq' an attribute is defined which associates a natural number with each object of these both sorts, and which is named by 't-attr' and 'seq-attr' respectively.

The action of the combine operator can be pictorially described as follows:

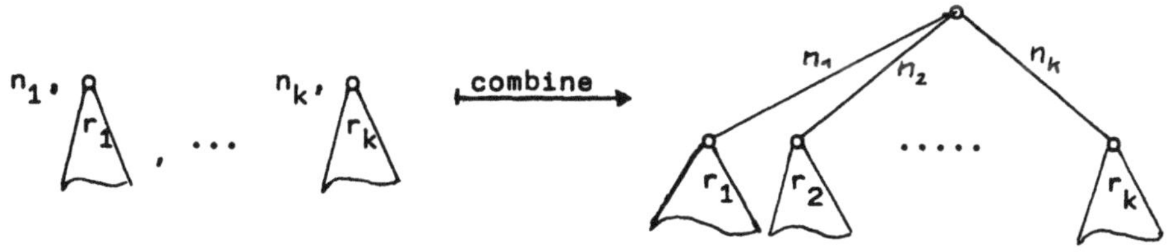

The following pictorially represented tree is not an object of sort 'Tree' since the leftmost path induces the sequence 3,5 which is not a decreasing sequence:

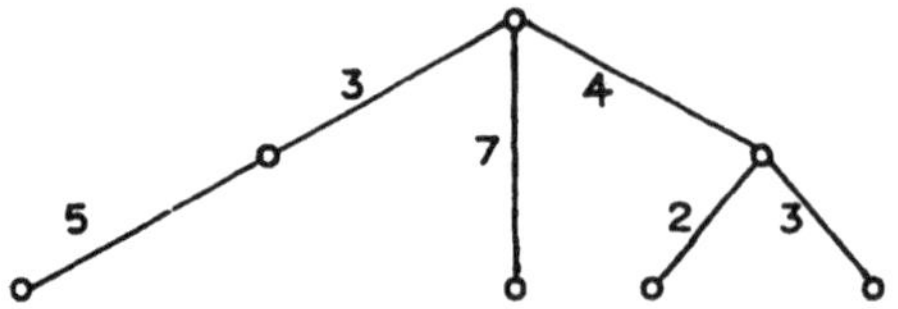

If we use the abbreviation r_o = combine(nil), the formal expression r= combine(add(add(add(nil,r_o,3),combine(add(add(nil,r_o,1),r_o,3)),4),r_o,2)) corresponds to the following pictorial representation:

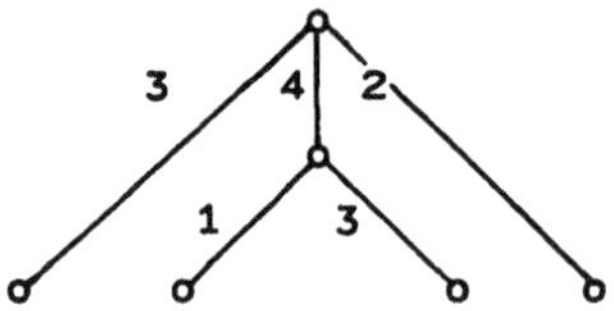

= r .

Furthermore the following equations hold:

t-attr(r) = 4

t-attr(combine(add(add(nil,r_o,1),r_o,3))) = 3 .

Now it seems to be necessary to say some words concerning the partiality. We have associated with the add-operator a necessary and sufficient domain condition. The necessity of the domain condition restricts the class of all possible interpretations in the intended way. In the preceding specification exactly the necessity causes that only decreasing sequences of edge values are generated by any path from the root to any leaf. In comparison to attribute grammars the necessity of the domain condition represents the context condition. However, if we would make available only tools for restrictions of the applicability of operations, then the concept of initial partial algebras would not work in the expected way. For partial algebras the initiality has an additional effect. Initiality causes minimal domains of partial operations. With respect to the specification of decreasing edge-valued trees this means that the domain of the add-operator would become empty if the domain condition would only be necessary. In this case 'nil' would be the only object of sort 'Treeseq' and 'combine(nil)' would be the only object of sort 'Tree'. Now we see that the sufficiency of domain conditions of partial operators is the tool to overcome this weakness of ordinary initial partial algebras. Thus, necessary and sufficient domain conditions of partial operators are the proper algebraic counterpart to context conditions in attribute grammars, because a production rule is applicable if and only if the context condition is satisfied.

As next we will informally discuss the validity of equations and of conditional equations for partial algebras. A mathematically precise discussion of this topic is given in Chapter 2.

The left-hand sides and the right-hand sides of equations are built up from variables and operators as formal expressions in the same way as expressions in programming languages. If the expressions of an equation do not contain partial operators, then there is no problem. An interpretation of the sort names and of the operators, which is compatible with the declarations of the operators, delivers an algebra A . A valuation of the variables of an equation in an algebra A is called a solution of the equation in A , if the valuation of the left-hand expression and of the right-hand expression evaluates to the same value. An equation is satisfied in an algebra A if each possible valuation of the variables is a solution of that equation in A . If the left-hand expression of the right-hand expression contains one or more partial operators, then these expressions are not evaluable for each valuation of the variables. In this case a valuation of the variables in a partial algebra A is said to be a solution in A if both sides of the equation are defined and if the resulting values are equal. Straightforwardly we define the set of solutions of a set of equations (all have to be defined over the same system of variables) as the intersection of the set of solutions of the single equations. Since a conditional equation is of the form

$$\underline{\text{if}}\ G\ \underline{\text{then}}\ e_1 = e_2$$

where G is a finite, possibly empty set of equations, we say that a partial algebra A satisfies that conditional equation if each solution of G in A is a solution of $e_1 = e_2$ in A, too. For reasons of convenience we define that any valuation of the variables in a partial algebra A is a solution of the empty set of equations. Applying this to a conditional equation with an empty premise, i.e. to

$$\underline{\text{if}}\ \emptyset\ \underline{\text{then}}\ e_1 = e_2 \quad ,$$

we obtain that a partial algebra satisfies that conditional equation iff e_1 and e_2 both are evaluable for each valuation of the variables and if equal values are produced. According to this fact, we may consider equations as abbreviated notations of conditional equations with an empty premise. This implies that all axioms in a specification are conditional equations, but some of them may have a trivial, i.e. an empty premise.

There is another effect deserving attention. Conditional equations cause applicability requirements. We illustrate this effect by the axiom

$$\underline{\text{if}}\ \ x \notin r\ \ \underline{\text{then}}\ \ y \in \text{add}(x,r) = \text{eq}(x,y)$$

in EXAMPLE1. The only partial operator in this specification is the add-operator. Bearing in mind the domain condition of the add-operator the premise says that add(x,r) is defined. If we would omit the premise, or more precisely, if we would substitute the premise $x \notin r$ by the empty set of equations, then the axiom

$$\underline{\text{if}}\ \ \emptyset\ \ \underline{\text{then}}\ \ y \in \text{add}(x,r) = \text{eq}(x,y)$$

would cause that the add-operator would become a total operator, and this would semantically contradict the intention of the domain condition. But this effect of conditional equations opens another style of specifications. One can omit the domain condition in the declaration of the operator and can achieve its effect by two additional axioms. In this way we would obtain the following specification EXAMPLE1* which is equivalent to the previous specification EXAMPLE1.

EXAMPLE1* <u>is</u> NAT <u>with</u> <u>definition</u>
    <u>sorts</u> Rseq
    <u>oprn</u> nil $\longrightarrow$ Rseq
        Nat $\in$ Rseq $\longrightarrow$ Bool
        add(Nat,Rseq) $\multimap\!\!\!\rightarrow$ Rseq
    <u>axioms</u> x,y: Nat, r: Rseq
    <u>if</u> $x \notin r$ <u>then</u> add(x,r) = add(x,r)
    <u>if</u> add(x,r) = add(x,r) <u>then</u> $x \notin r$
    $x \notin$ nil
    <u>if</u> $x \notin r$ <u>then</u> $y \in$ add(x,r) = eq(x,y)
<u>end</u> EXAMPLE1*

Now the first axiom expresses the sufficiency of the domain condition and the second axiom expresses the necessity.

We do not prefer one of these styles of specifications. If the domain condition is simple we prefer the specification of the domain condition within the declaration of the operator just like in EXAMPLE1 . Unfortunately it is not a simple task to check whether the axioms cause additional applicability requirements. A corresponding criterion will be discussed in Section 2.5. But we will also see examples for which a specification in the style of EXAMPLE1* seems to be more convenient. The specification of the semantics of a 'while...do...-statement' is such an example. This specification will be given in the following section.

In a specification of an ADT one should use minimal systems of operators for improving the readability. To achieve this simplicity one should specify only those operators that are necessary for the generation of all objects of the intended data type, and that are necessary for the domain conditions and for the specification of the intended properties of interrelations between the generating operators. In this sense the specification of decreasing edge-valued trees is minimal. If we would apply this principle to natural numbers and truth values, we would only introduce the constants 'zero', 'true', and 'false' and the unary successor operator. But there are different points of views of assessing specifications. The suggested purity of a specification is advantageous concerning the understandability, but it is evidently a disadvantage concerning the value in use. The improvement of the value in use of an ADT generally requires functional enrichments of a basically specified ADT.

According to the preceding consideration we discuss functional enrichments of ADT's in the following.

First of all, for each ADT the specification of the identity should be added. Therefore we start with the specification of identities of restricted sequences of natural numbers and of decreasing edge-valued trees.

The main problem in functional enrichments of ADT's is the consistency and persistency of the new operators and new axioms with the given ADT. This means that initial satisfaction of the enriched specification does not cause identifications of existing objects of the given ADT, and that by means of the added operators no new objects will be created, and finally that the domain of given operators will not be changed.

The following enrichment of EXAMPLE1 does not fulfil the requirements of consistency and persistency:

EXAMPLE3 <u>is</u> EXAMPLE1 <u>with</u> <u>definition</u>
        <u>oprn</u> r-eq(Rseq,Rseq) ⟶ Bool
        <u>axioms</u> m: Rseq
        r-eq(m,m) = true
  <u>end</u> EXAMPLE3

Because of initiality the new operator 'r-eq' generates for each pair (m_1,m_2) of different restricted sequences a new object of sort 'Bool'. Thus, the given enrichment does not preserve the semantics of enriched specification. The generation of new objects is a consequence of the absence of axioms that force $r\text{-}eq(m_1,m_2)$ = false for each pair (m_1,m_2) with $m_1 \neq m_2$. But this statement of the intended property of the enriched operator 'r-eq' uses the identity of restricted sequences which we just want to specify. To break through this circle we go back to the inductive generation of restricted sequences and try to specify inductively the intended semantics of 'r-eq'.

EXAMPLE4 is EXAMPLE1 with definition
    oprn r-eq(Rseq,Rseq) $\longrightarrow$ Bool
    axioms x,y: Nat, m,m_1,m_2: Rseq
    r-eq(m,m) = true
    if $x \notin m_1$, $y \notin m_2$ then
        r-eq(add(x,m_1),add(y,m_2)) = and(eq(x,y),r-eq(m_1,m_2)) fi
end EXAMPLE4

The specification EXAMPLE4 seems to be an inductive specification of the intended equality. A more careful study points to the open problem of checking the equality of restricted sequences of different length. For instance, there is no way of proving only by means of the given two axioms that r-eq(nil,add(x,nil)) is equal to true or to false. This lack of a formal proof causes again by initiality the creation of new objects of sort 'Bool'. The failure in the preceding specification could be corrected by an additional specification of a length function and by adding an axiom that causes the inequality of restricted sequences of different length according to the following specification.

EXAMPLE5 is EXAMPLE1 with definition
    oprn length(Rseq) $\longrightarrow$ Nat
    axioms x: Nat, r: Rseq
    length(nil) = zero
    if $x \notin r$ then length(add(x,r)) = succ(length(r))
    with definition
    oprn r-eq(Rseq,Rseq) $\longrightarrow$ Bool
    axioms x,y: Nat, m,m_1,m_2: Rseq
    r-eq(m,m) = true
    if $x \notin m_1$, $y \notin m_2$ then
        r-eq(add(x,m_1),add(y,m_2)) = and(eq(x,y),r-eq(m_1,m_2)) fi
    if eq(length(m_1),length(m_2))=false then r-eq(m_1,m_2)=false
end EXAMPLE5

EXAMPLE5 is not aimed to give a most elegant specification of the equality of restricted sequences of natural numbers. The main object was the demonstration of problems that appear in connection with functional enrichments of ADT's. But, we have not yet discussed all kinds of problems. We have only dealed with creation of new objects as a consequence of insufficient systems of axioms.

In the following we will discuss problems that are based on the partiality of generating operators. For that purpose we consider the specification EXAMPLE5* that results from EXAMPLE5 by removing the condition $y \notin m_2$ from the premise of the last but one axiom. This removal causes the totality of the add-operator, since for each assignment of the variables x,y: Nat, m_1,m_2: Rseq with $x \notin m_1$ the formal expression $r\text{-}eq(add(x,m_1),add(y,m_2))$ becomes evaluable and so the subexpression $add(y,m_2)$ becomes evaluable too for such an assignment. But now there is no precondition for the addition of y to m_2.

The preceding discussion shows that the partiality of generating operators does not seriously complicate inductive specifications of functional enrichments. The only thing that has to be born in mind is, that the premise has to contain all those conditions that guarantee the evaluability of all expressions and subexpressions of the conclusion.

The third type of incompatibility of an enrichment with a given specification concerns the identification of objects. If we would substitute in EXAMPLE5 the axiom

<u>if</u> $x \notin m_1,\ y \notin m_2$ <u>then</u>

$r\text{-}eq(add(x,m_1),add(y,m_2)) = and(eq(x,y),r\text{-}eq(m_1,m_2))$ <u>fi</u>

by the following axiom

<u>if</u> $x \notin m_1,\ y \notin m_2$ <u>then</u>

$r\text{-}eq(add(x,m_1),add(y,m_2)) = or(eq(x,y),r\text{-}eq(m_1,m_2))$ <u>fi</u>

the resulting specification would cause the identification of the both truth values. Because of the modified axiom we would obtain for $m_1 = add(z,nil)$, $m_2 = nil$ the equality

$true = or(eq(x,x),r\text{-}eq(m_1,m_2) = r\text{-}eq(add(x,m_1),add(x,m_2))$, and because of $length(add(x,m_1)) = 2 \neq 1 = length(add(x,m_2))$ the last axiom implies $r\text{-}eq(add(x,m_1),add(x,m_2)) = false$.

As next we specify the equality of decreasing edge-valued trees without an extensive failure discussion.

DECREASING EDGE-VALUED TREES WITH EQUALITY <u>is</u> DECREASING EDGE-VALUED TREES <u>with</u> <u>definition</u>

<u>oprn</u> t-eq(Tree,Tree) ⟶ Bool

s-eq(Treeseq,Treeseq) ⟶ Bool

Bool ∧ Bool ⟶ Bool

<u>axioms</u> x: Bool, s_1,s_2,s: Treeseq, r_1,r_2,r: Tree, n_1,n_2: Nat

$false \wedge x = x \wedge false = false$

$true \wedge true = true$

$t\text{-}eq(combine(s_1),combine(s_2)) = s\text{-}eq(s_1,s_2)$

$s\text{-}eq(nil,nil) = true$

$s\text{-}eq(s_1,s_2) = s\text{-}eq(s_2,s_1)$

$\underline{if}$ $t\text{-}attr(r) < n$ $\underline{then}$ $s\text{-}eq(add(s,r,n),nil) = false$

$\underline{if}$ $t\text{-}attr(r_1) < n_1,\ t\text{-}attr(r_2) \quad n_2$ $\underline{then}$

$s\text{-}eq(add(s_1,r_1,n_1)),add(s_2,r_2,n_2)) =$

$= (eq(n_1,n_2) \wedge t\text{-}eq(r_1,r_2)) \wedge s\text{-}eq(s_1,s_2)$ $\underline{fi}$

__end__ DECREASING EDGE-VALUED TREES WITH EQUALITY

Even if in many papers on abstract data types it is assumed that for each sort the equality is available, the equality is not initially specifiable for each sort in each ADT. The undecidability of the word problem in semigroups and other examples of recursively undecidable problems allow the specification of ADT's where the equality is not recursively decidable so that according to Section 4.2 the equality is also not initially specifiable. Later we will see that for ADT's with partial operations the initial specifiability of the equality is equivalent to an interesting topic namely to the recursiveness of an ADT. Informally we call an ADT recursive if the isomorphism class, which semantically represents the ADT , contains an algebra with recursively enumerable carrier sets and with partial recursive fundamental operations, i.e. if there is a computable prototype of the given ADT . Unfortunately it is recursively undecidable whether an arbitrary specification has a computable prototype or not.

We will conclude this section with two further enrichments of DECREASING EDGE-VALUED TREES . By the first functional enrichment we specify a function that associates with an object t of sort 'Tree' a natural number $depth(t)$ that gives the maximal length of all paths over t from the root to leaves. By the second functional enrichment we specify a function that gives for an object t of sort 'Tree' the sum over all edge values of t . For the following graphically represented decreasing edge-valued tree t_o then hold $depth(t_o) = 3$ and $sum(t_o) = 18$.

t_o =

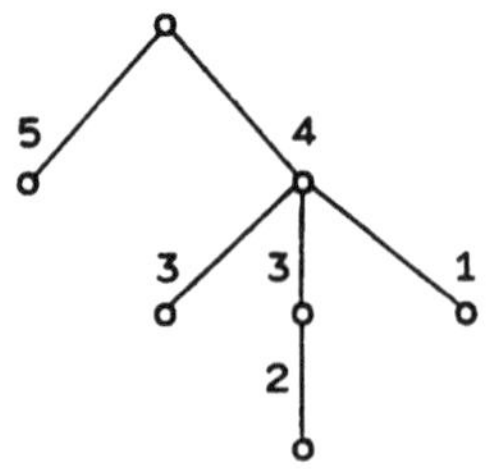

```
EXAMPLE6 is DECREASING EDGE-VALUED TREES with definition
        oprn  depth(Tree) ──→ Nat
              s-depth(Treeseq) ──→ Nat
        axioms  n: Nat, r: Tree, s: Treeseq
        depth(combine(nil)) = zero
        s-depth(nil) = zero
        if t-attr(r) < n  then
                    depth(combine(add(s,r,n)))=succ(s-depth(s)) fi
        if  t-attr(r) < n  then
                    s-depth(add(s,r,n)) = max(s-depth(s),depth(r))  fi
  end EXAMPLE6
```

This style of defining a function over an ADT seems to be very natural. We have only to think about how to compute the intended value if we know the corresponding values of all constituents with respect to all generating operators and how to compute the intended value for basic generating objects. Since not all constituents must be of the same sort it may happen that a simultaneous specification of an auxiliary operator is very helpful if not even necessary. In the preceding specification the operator 's-depth' serves as such an auxiliary function. In the same manner we can specify the second function.

```
EXAMPLE7 is DECREASING EDGE-VALUED TREES with definition
        oprn  Nat + Nat ──→ Nat
              sum(Tree) ──→ Nat
              s-sum(Treeseq) ──→ Nat
        axioms  x,y: Nat, r: Tree, s: Treeseq
        zero + y = y
        succ(x) + y = succ(x + y)
        sum(combine(s)) = s-sum(s)
        s-sum(nil) = zero
        if t-attr(r) < x then
                    s-sum(add(s,r,x)) = (s-sum(s) + sum(r)) + x  fi
 end EXAMPLE7
```

In EXAMPLE7 we specified at first the addition of natural numbers by the well-known inductive definition. The same principle is used for the specification of the sum-operator and of the auxiliary operator s-sum.

Finally we point out that the demonstrated style of defining a function over an ADT significantly differs from the imperative style as used in ordinary programming languages. In terms of an imperative programming language it would not be possible to define one of the functions depth(Tree) ⟶ Nat and sum(Tree) ⟶ Nat using only the functions made available by the specification DECREASING EDGE-VALUED TREES because of the lack of operators which decompose compound objects.

1.2 Parameterized Abstract Data Types - PADT's

The examples of the previous section are so simple that formal specifications are not necessary for dealing with such objects. Formal specifications are destined for complex problems for which no suitable mathematical models exist. However, it is an illusion to expect that specifications of complex problems have to be simple. First experiences with formal specifications show that formal specifications may also become complex and therefore hard to read, hard to write, and hard to modify as a whole.

The main advantage of formal specification methods is the possibility of breaking down the complex problem into well defined pieces of understandable size. It was the specification language CLEAR, invented by Burstall and Goguen (1977) that put in the centre of interest facilities for writing structured specifications. By means of parameterized specifications a large real-world problem can be broken down and built up from small, easy to understand reusable pieces.

The problem of structuring large specifications is one of the central research problems in program methodology. A detailed discussion of structuring principles or a review of existing approaches to structures of specifications is out of the range and out of the aim of this book. In the opinion of the author there are at least two concepts that should be generally accepted, namely the 'CAT-concept' of Goguen and Burstall (1980) and the idea of 'data institutions' of Burstall and Goguen (1982).

The CAT-concept reflects the important fact that structuring of specifications is only one dimension of the whole problem. Formal specification methods should not only be used for specifications of the functional behaviour of a large program before writing that program in a programming language. If the conceptual distance between the basic notions of the specification and the basic notions of the programming language is very large, intermediate conceptual levels are necessary. Formal specifications should also be used for descriptions of transformations of the basic notions of one conceptual level into compound notions of the next lower conceptual level. The composition of specifications that represent transformations in the direction from the highest conceptual level to the conceptual level of the used programming language gives the second dimension of the complexity of a large program. In this way two kinds of compositions of specifications exist. Goguen and Burstall (1980) call these compositions the horizontal and the vertical composition.

Data Institutions are aimed to give a mathematical foundation that allows the definition and investigation of specification-building operations largely independent of the concrete mathematical semantics that is used for the specifications of ADT . In the terminology of data institutions it is the aim of this book to give an introduction, an investigation, and an illustration of a special data institution that we will call 'Initial Canons', according to the report of Hupbach, Kaphengst, and Reichel (1980). As mentioned earlier the main characteristic of initial canons is the availability of partial operations.

Before we generally explain the facilities for writing structured initial canons we will give a simple example of a parameterized abstract data type. We specify the PADT of finite directed paths over directed graphs. As distinct from specifications of ADT's we have now not only to specify an isomorphism class of a many-sorted algebra. In the following example we have at first to specify the class of directed graphs which act as parameters. On the basis of that specification of the parameter class we have to specify the uniform construction of the set of directed paths over an arbitrary directed graph.

```
DIRECTED GRAPH is requirement
        sorts  Edges, Nodes
        oprn   begin(Edges) ——→ Nodes
               end(Edges) ——→ Nodes
  end DIRECTED GRAPH
```

The key word <u>requirement</u> indicates that interpretations of the following sort names and of the following operators are not restricted to initial satisfaction . Since DIRECTED GRAPH does not contain any axiom, each two-sorted algebra

$$G = (G_{Edges}, G_{Nodes}; \; begin^G : G_{Edges} \longrightarrow G_{Nodes}, end^G : G_{Edges} \longrightarrow G_{Nodes})$$

is a model of the specification DIRECTED GRAPH , where G_{Edges} and G_{Nodes} are the sets of edges and nodes of G , respectively, and an edge $x \in G_{Edges}$ is directed from $begin^G(x) \in G_{Nodes}$ to $end^G(x) \in G_{Nodes}$.

By the next step we specify the parameterized construction of the set of finite directed paths.

FINITE PATHS <u>is</u> DIRECTED GRAPH <u>with</u> <u>definition</u>

<u>sorts</u> Path

<u>oprn</u> nil(Nodes) ⟶ Path

add(w:Path,x:Edges <u>iff</u> p-end(w)=begin(x)) ⟶ Path

p-end(Path) ⟶ Nodes

<u>axioms</u> e: Nodes, w_1,w_2: Path, x: Edges

p-end(nil(e)) = e

<u>if</u> w_2 = add(w_1,x) <u>then</u> p-end(w_2) = end(x)

<u>end</u> FINITE PATHS

If we want to extend an interpretation G of DIRECTED GRAPH to an interpretation W(G) of the extended initial canon FINITE PATHS the key word <u>definition</u> requires that W(G) is an initial extension of G . This means

(i) The given interpretation G has to be preserved by the initial extension W(G) , i.e.

$$W(G)_{Edges} = G_{Edges},\quad W(G)_{Nodes} = G_{Nodes},$$

$$begin^{W(G)} = begin^G,\quad end^{W(G)} = end^G .$$

(ii) The interpretation of the additional sort names and operators is restricted by initial **satisfaction** .

We call the condition (i) the persistency and the condition (ii) the initiality of an enrichment by definition of a canon.

It may happen that persistency and initiality together prevent the existence of any initial extension. In this case the initial canon semantically represents the empty class of algebras and is therefore of no practical interest. Since there are no syntactic rules that prevent the writing of initial canons with empty model classes, we have to bear in mind this pathological case, too.

We illustrate the persistency and initiality of an enrichment by definition of a canon by the following incorrect enrichment.

FINITE PATHS* <u>is</u> DIRECTED GRAPH <u>with</u> <u>definition</u>

<u>sorts</u> Path

<u>oprn</u> nil(Nodes) ⟶ Path

add(w:Path,x:Edges <u>iff</u> p-end(w)=begin(x)) ⟶ Path

p-end(Path) ⟶ Path

<u>axioms</u> w_1,w_2: Path, x: Edges

if w_2 = add(w_1,x) **then** p-end(w_2) = end(x)

end FINITE PATHS*

Both initial canons differ only in the set of axioms. If G is any directed graph, then there is up to isomorphism exactly one extension W(G) that interprets the additional sort names and operators, and that initially satisfies the additional axiom of FINITE PATHS* . But this unique extension W(G) is not a persistent extension of G . At first we obtain exactly one object $nil^{W(G)}(e)$ in $W(G)_{Path}$ for each node e in G_{Nodes} . As next we can apply the operation $p\text{-}end^{W(G)}$ to each of these new object. Initiality causes that in this way new nodes are created, i.e. $W(G)_{Nodes} \supsetneq G_{Nodes}$. This generation of additional nodes is prevented in the specification FINITE PATHS by the axiom p-end(nil(e)) = e . Additionally the generation of new nodes implies that the domain of $add^{W(G)}$ becomes empty since there is no edge x in $W(G)_{Edges}$ with

$$begin^{W(G)}(e) = p\text{-}end^{W(G)}(nil^{W(G)}(e))$$

for any e in $W(G)_{Nodes}$. Now the operation $nil^{W(G)}$ can be applied to the created new nodes. This application produces new object in $W(G)_{Path}$ which can be used as arguments of the operation $p\text{-}end^{W(G)}$. Applying this operation in this way one produces again new nodes in $W(G)_{Nodes}$. This process can be applied in an unrestricted number of steps. The resulting set $W(G)_{Nodes}$ properly includes the set of nodes of the given directed graph. The preceding consideration proves that there is no interpretation of FINITE PATHS* that satisfies both the persistency and the initiality.

After the discussion of possible conflicts between persistency and initiality we come back to the specification FINITE PATHS . As previously mentioned the creation of additional nodes is prevented by the first axiom. As before, G may denote an arbitrary directed graph. If there are an edge $x_1 \in G_{Edges}$ and a node $e \in G_{Nodes}$ with $e = begin^G(x_1)$, then the operation $add^{W(G)}$ is applicable to the pair $(nil^{W(G)}(e),x_1)$ and an object of sort 'Path' is produced as value of the formal expression add(nil(e),x_1). If it is evident from context in which algebra we are working, we allow to omit the name of the algebra in the notation of an operation. According to that agreement in the given context add(nil(e),x_1) abbreviates $add^{W(G)}(nil^{W(G)}(e),x_1)$. If in G_{Edges} there is an edge x_2 not necessarily different from x_1 but satisfying end(x_1) = begin(x_2) , then the operation add is applicable to (add(nil(e),x_1),x_2) and a further path as value of the expression add(add(nil(e),x_1),x_2) will be produced. The applicability of add to $(add(nil(e),x_1),x_2)$ is a consequence of the second axiom. By setting $w_2 = add(nil(e),x_1)$ and $w_1 = nil(e)$ we obtain $p\text{-}end(add(nil(e),x_1)) = end(x_1)$ and because of $end(x_1) = begin(x_2)$ we finally get $p\text{-}end(add(nil(e),x_1)) = begin(x_2)$. The expression $add(add(nil(e),x_1),x_2)$ represents a sequence of two connected edges or in other words a path of length two from $begin(x_1) = e$ to $end(x_2)$. The sufficiency of the domain condition of the add-operator guarantees that each finite directed path will be created, whereas the necessity of the domain condition prevents the construction of non-connected sequences of edges.

The initial canon FINITE PATHS contains only those operators that are necessary for the generation of all finite directed paths. For working with this PADT it would be convenient to enrich this PADT, for instance with a function that associates with each path its source, or with a function that calculates the length of each path, or one can define the partial composition of paths that concatenates two paths, iff they are connected. The following initial canon is a stepwise enrichment of FINITE PATHS by specifications of the suggested functions.

```
EXAMPLE2.1 is  FINITE PATHS  with  NAT  with definition
        oprn  length(Path)  ──→ Nat
        axioms  x: Edges,  w1,w2: Path,  e: Nodes
        length(nil(e)) = zero
        if  w2 = add(w1,x)  then  length(w2) = succ(length(w1))
        with definition
        oprn  source(Path)  ──→ Nodes
        axioms  x: Edges,  w1,w2: Path,  e: Nodes
        source(nil(e)) = e
        if  w2 = add(w1,x)  then  source(w2) = source(w1)
        with definition
        oprn compose(w1:Path,w2:Path iff p-end(w1)=source(w2)) —
                                                       ──→ Path
        axioms  w1,w2: Path,  x: Edges
        compose(w1,nil(p-end(w1))) = w1
        if  p-end(w1) = source(add(w2,x))  then
                      compose(w1,add(w2,x))=add(compose(w1,w2),x) fi
  end EXAMPLE2.1
```

In the first line we joint by the key word with the initial canon FINITE PATHS with the initial canon NAT since we need natural numbers as results of the required function length(Path) ──→ Nat. The principle we used for the unique specifications of the functions length, source, and composition is evidently the same we used in Section 1.1 for functional enrichments of ADT's. But now the operators do not represent an abstract algorithm over an ADT, now the operators represent parameterized abstract algorithms which become abstract algorithms if the parameter structure is actualized by an ADT which specifies a directed graph. The problem of parameter substitution will be discussed in detail at the end of this section in connection with a further structuring principle of initial canons called 'modification'.

The possibility of combinations of enrichments by definitions and enrichments by requirements significantly increases the expressiveness of initial canons. We demonstrate this mixed enrichments by the following specification. At first we enrich FINITE PATHS by a requirement of an integer valuation of edges, and in a second enrichment by definition we use this valuation for the definition of a function that gives for each path the minimal valuation over all edges of that path.

```
EXAMPLE2.2  is  FINITE PATHS  with  NAT  with requirement
        oprn  capacity(Edges)  ──▶  Nat
        with definition
        oprn   min(Nat,Nat)  ──▶  Nat
               Nat < Nat   ──▶  Nat
        axioms  x,y: Nat
        x ≮ zero
        zero < succ(x)
        (x < y) = (succ(x) < succ(y))
         min(x,x) = x
         if  x < y  then  min(x,y) = x
         with definition
         oprn  cap(Path)  ──▶  Nat
         axioms  e: Nodes, x,y: Edges,  w1,w2: Path
         cap(nil(e)) = zero
         if e=begin(x) then  cap(add(nil(e),x))=capacity(x)
         if  w2 = add(add(w1,x),y)  then
                   cap(w2) = min(cap(add(w1,x)),capacity(y))  fi
  end EXAMPLE2.2
```

This example makes evident the difference between both kinds of enrichments of initial canons. On a more abstract level the difference is given by the fact that enrichments by definition are characterized by persistency and initiality whereas for enrichments by requirement only persistency is claimed.

In Section 1.1 the specification of the identity of an ADT was not a trivial task. The strong expressiveness of initial canons makes this problem to a trivial one as we can see by the following specification.

```
EXAMPLE2.3  is FINITE PATHS  with  BOOL  with requirement
        oprn  equal(Path,Path) ——> Bool
        axioms  w,w1,w2: Path
        equal(w,w) = true
        if  equal(w1,w2) = true  then  w1 = w2
  end EXAMPLE2.3
```

This specification of the identity of paths is not based on the inductive definition of finite paths. Even if the operator equal is introduced within an enrichment by requirement, the persistency together with the stated axioms guarantees that this operator can be interpreted only in one way, namely as the intended identity. In this way we can semantically uniquely specify the identity for each sort name of an initial canon.

The equivalent problem to Section 1.1 is now the question whether the identity of paths can be specified by an enrichment by definition of FINITE PATHS. Intuitively it seems to be clear that the answer is no, since there is no possibility to check the identity of edges. By means of results of Section 4.2 we can confirm this intuition. But, if the identity of edges is available, it is not hard to find an inductive definition of the identity of paths. There are several possibilities for such a specification. One of them is the following.

```
EXAMPLE2.4  is  FINITE PATHS  with  BOOL  with requirement
        oprn  e-eq(Edges,Edges) ——> Bool
        axioms  x,y: Edges
        e-eq(x,x) = true
        if  e-eq(x,y) = true  then   x = y
        with definition
        oprn  equal(Path,Path) ——> Bool
        axioms  e: Nodes, x,y: Edges, u1,u2,w1,w1: Path
        equal(w1,w1) = true
        equal(w1,w2) = equal(w2,w1)
        if p-end(w1)=begin(x) then  equal(w1,add(w1,x))=false
        if  u1 = add(w1,x),  u2 = add(w2,y)  then
                 equal(u1,u2)=(equal(w1,w2) ∧ e-eq(x,y))  fi
  end EXAMPLE2.4
```

In connection with the substitution of parameters it is convenient to assume that for each sort name of an initial canon that is introduced within an enrichment by requirement also its **equality** operator is introduced. Since those sort names represent parameters, this assumption corresponds to the availability of the identity for each parameter type. One intended consequence of this assumption is, that those parameter types can be actualized only by computable ADT's, because the required identity is then also subject to actualization. According to this convention we enrich DIRECTED GRAPH to D-GRAPH:

D-GRAPH <u>is</u> DIRECTED GRAPH <u>with</u> BOOL <u>with</u> <u>requirement</u>
  <u>oprn</u> n-eq(Nodes,Nodes) ⟶ Bool
    e-eq(Edges,Nodes) ⟶ Bool
  <u>axioms</u> x,y: Nodes, a,b: Edges
  n-eq(x,x) = true
  <u>if</u> n-eq(x,y) = true <u>then</u> x = y
  e-eq(a,a) = true
  <u>if</u> e-eq(a,b) = true <u>then</u> a = b
<u>end</u> D-GRAPH

In the preceding specifications BOOL denotes a specification of the ADT of truth values with the usual operations $\wedge$, $\vee$, $\neg$.

As next we study the limits of expressiveness of enrichments by definitions of initial canons. In Section 4.2 we will show that it is justified to consider enrichments by definitions that define functional extensions as specifications of computable functions that are parameterized by the given initial canon, or as parameterized computable function for short. Let the given base canon be D-GRAPH and let us consider the parameterized function

$$\text{connected}^G(G_{\text{Nodes}}, G_{\text{Nodes}}) \longrightarrow \{\text{true}, \text{false}\}$$

where G denotes any model of D-GRAPH . This function may take the value true for an ordered pair (e_1, e_2) of nodes in G if and only if there is a finite directed path over G from e_1 to e_2. Now we can put the question **whether this parameterized function** is computable over D-GRAPH or not.

In Chapter 4 we prove properties of enrichments by definitions that give an answer to this question. Especially Theorem 4.2.4 gives a necessary condition for computable functions parameterized by an initial canon $\mathbb{B}$. Roughly speaking this theorem states the compatibility of homomorphisms between models of $\mathbb{B}$ with the considered function. With respect to our example this means that for each homomorphism $f: A \longrightarrow B$

between models A and B of D-GRAPH it holds

$$connected^A(e_1,e_2) = connected^B(f_{Nodes}(e_1),f_{Nodes}(e_2))$$

for all $e_1,e_2 \in A_{Nodes}$, if the function 'connected' is computable with respect to D-GRAPH .

Since a homomorphism $f: A \longrightarrow B$ between models of D-GRAPH has to be compatible with each operator of D-GRAPH (see Section 2.2), the single functions

$$f_{Nodes}: A_{Nodes} \longrightarrow B_{Nodes} \ , \ f_{Edges}: A_{Edges} \longrightarrow B_{Edges}$$

satisfy the following conditions:

$f_{Nodes}(begin^A(x)) = begin^B(f_{Edges}(x))$ and

$f_{Nodes}(end^A(x)) = end^B(f_{Edges}(x))$ for each $x \in A_{Edges}$, and

$n\text{-}eq^A(e_1,e_2) = n\text{-}eq^B(f_{Nodes}(e_1),f_{Nodes}(e_2))$ for all e_1,e_2 of A_{Nodes},

$e\text{-}eq^A(x_1,x_2) = e\text{-}eq^B(f_{Edges}(x_1),f_{Edges}(x_2))$ for all edges $x_1,x_2 \in A_{Edges}$.

These conditions imply that for each homomorphism $f: A \longrightarrow B$ between models of D-GRAPH the functions $f_{Nodes}: A_{Nodes} \longrightarrow B_{Nodes}$ and $f_{Edges}: A_{Edges} \longrightarrow B_{Edges}$ are one-to-one.

For the given example it is now easy to find models A and B of D-GRAPH and to construct a homomorphism $f: A \longrightarrow B$ that is not compatible with the parameterized function

$$connected(Nodes,Nodes) \longrightarrow Bool \ .$$

We can take models A and B of D-GRAPH with $A_{Nodes} = B_{Nodes} = M$ an arbitrary set, with $A_{Edges} = \emptyset$ the empty set, with B_{Edges} any non-empty set, and with any functions $begin^B$, end^B.
Evidently we set $A_{Bool} = B_{Bool} = \{true, false\}$.
With this assumption we obtain a homomorphism $f: A \longrightarrow B$ by setting $f_{Nodes} = Id_M$ and taking $f_{Edges}: \emptyset \longrightarrow B_{Edges}$ the unique inclusion of the empty set, and by $f_{Bool} = Id_{\{true,false\}}$.
Because of $A_{Edges} = \emptyset$ any two nodes of A are inconnected . Now we can choose B_{Edges} , $begin^B$, and end^B in such a way that there are at least two $e_1,e_2 \in B_{Nodes} = A_{Nodes}$ that are connected in B by a finite directed path from e_1 to e_2 . For such a choice we obtain

$connected^A(e_1,e_2) = false$

$connected^B(f_{Nodes}(e_1),f_{Nodes}(e_2)) = connected^B(e_1,e_2) = true$,

so that this homomorphism is not compatible with the considered function. This shows that the class of models of D-GRAPH is much to rich to allow a definition of the intended paramaterized function, or in other words, there are not enough fundamental operations for a definition of the connected-operator.

One can summarize the preceding considerations in the phrase that the computational capacity of the initial canon D-GRAPH is too weak for a specification of the intended function by means of enrichments by definitions. Perhaps this result is not too surprising because we permit any directed graph as parameter. What about the restriction to directed graphs with finite sets of nodes and edges? Would this restriction sufficiently increase the computational capacity? Before we try to answer this question we have to answer another one. Is it possible to represent the class of all finite directed graphs as the class of all models of a suitable initial canon? Only if the answer to that question would be yes the question before becomes meaningful. By the following specification we demonstrate that the class of all finite directed graphs is representable as class of models of an initial canon, although that class is out of range of first order logic.

```
FINITE DIRECTED GRAPH  is  D-GRAPH  with  definition
       sorts  Set
       oprn   empty ——> Set
              e-add(Set,Edges) ——> Set
              n-add(Set,Nodes) ——> Set
       axioms  a,a1,a2: Nodes,  x,x1,x2: Edges,  s: Set
       e-add(e-add(s,x),x) = e-add(s,x)
       n-add(n-add(s,a),a) = n-add(s,a)
       e-add(e-add(s,x1),x2) = e-add(e-add(s,x2),x1)
       n-add(n-add(s,a1),a2) = n-add(n-add(s,a2),a1)
       e-add(n-add(s,a),x) = n-add(e-add(s,x),a)
       with requirement
       oprn   edges ——> Set
              nodes ——> Set
       axioms  a: Nodes,  x: Edges
       e-add(edges,x) = edges
       n-add(nodes,a) = nodes
  end FINITE DIRECTED GRAPH
```

For each model A of FINITE DIRECTED GRAPH with $A_{Edges} \cap A_{Nodes} = \emptyset$ the set A_{Set} consists of all finite subsets of $A_{Edges} \cup A_{Nodes}$, $empty^A \in A_{Set}$ is the empty set, and the operations $e\text{-}add^A$, $n\text{-}add^A$ are given by

$$e\text{-}add^A(s,x) = s \cup \{x\} \text{ , and } n\text{-}add^A(s,a) = s \cup \{a\}$$

for each finite subset $s \in A_{Set}$ and each $x \in A_{Edges}$, $a \in A_{Nodes}$.
The assumed disjointness of A_{Edges} and A_{Nodes} is not a restriction of the generality. In this case A_{Set} becomes the set of all finite subsets of the disjoint union of A_{Edges} and A_{Nodes}. Due to the initiality and persistency of the enrichment by definition the interpretation of Set, empty, e-add, and n-add is unique up to isomorphism. However, the restriction to finite directed graphs does not result from the enrichment by definition but from the enrichment by requirement. This enrichment requires the distinction of two finite subsets

$$edges^A \in A_{Set}, \quad nodes^A \in A_{Set}$$

such that each node $a \in A_{Nodes}$ is an element of $nodes^A$, and each edge $x \in A_{Edges}$ is an element of $edges^A$. A distinction of those elements in A_{Set} is practicable only for a finite directed graph A.

After that explanation of the specification FINITE DIRECTED GRAPH we come back to the question whether the restriction to finite directed graphs sufficiently increases the computational capacity. However, the same construction of a homomorphism $f: A \longrightarrow B$ between directed graphs that has been used before can now also be used for an answer in the negative, since each homomorphism $f: A \longrightarrow B$ between finite directed graphs can uniquely extended to a homomorphism between models of FINITE DIRECTED GRAPH. This negative answer indicates that a sufficient increase of the computational capacity requires additional operations.

There are several ways for a sufficient increase of the computational capacity of D-GRAPH. But we will not attend to that matter further on, instead we will show that the intended function can be uniquely specified by a mixed enrichment.

C-D-GRAPH <u>is</u> D-GRAPH <u>with</u> <u>definition</u>

    <u>sorts</u> Path

    <u>oprn</u> nil(Nodes) $\longrightarrow$ Path

        add(w:Path,x:Edges <u>iff</u> p-end(w)=begin(x)) $\longrightarrow$ Path

        p-end(Path) $\longrightarrow$ Nodes

        p-begin(Path) $\longrightarrow$ Nodes

    <u>axioms</u> e: Nodes, x: Edges, w_1, w_2: Path

```
        p-end(nil(e)) = e
        p-begin(nil(e)) = e
        if  w2 = add(w1,x)  then  p-end(w2) = end(x)
        if  w2 = add(w1,x)  then  p-begin(w2) = p-begin(w1)
        with requirement
        oprn connected(Nodes,Nodes) ——▶ Bool
             path(x:Nodes,y:Nodes iff connected(x,y)=true) ——▶ Path
        axioms  x,y: Nodes,  w: Path
        if x=p-begin(w), y=p-end(w) then  connected(x,y) = true
        if connected(x,y) = true  then  x = p-begin(path(x,y))
        if connected(x,y) = true  then  y = p-end(path(x,y))
  end C-D-GRAPH
```

Both C-D-GRAPH and FINITE DIRECTED GRAPH clearly demonstrate the profound difference in the expressiveness of enrichments by definitions and of mixed **enrichments**.

The previous discussion was concerned with the limits of expressiveness of specifications. In the following we demonstrate the expressiveness of enrichments by definition if partial operations are available in contrast to specifications with total operations only. The following specification of the semantics of a loop-statement, for instance of the form

<u>while</u> p <u>do</u> g <u>else</u> f

will be used to prove the adequacy of structural induction on partial algebras. We assume that p: M —●▶ {true,false} is a partial predicate on a parameter set M , and that f,g: M —●▶ M are partial transformations.

```
WHILE  is BOOL  with requirement
       sorts  M
       oprn  p(M)  —●▶  Bool
             f(M)  —●▶  M
             g(M)  —●▶  M
       with definition
       oprn   while(M)  —●▶  M
       axioms  x: M
       if  p(x)=false, f(x)=f(x)  then  while(x) = f(x)
       if  p(x)=true, while(g(x))=while(g(x))  then
                      while(x) = while(g(x))  fi
  end  WHILE
```

For a correct understanding of this specification one should have in mind that applicability requirements can be expressed by axioms. In the specification **WHILE** the domain of the operator 'while' is defined by means of domain requirements resulting only from both axioms. The first axiom states that an element x is in the domain of 'while' whenever x is in the domain of p and in the domain of f and if additionally x does not satisfy p. In that case the first axiom additionally fixes the value $while(x) = f(x)$. The first axiom represents the basis of the inductive definition. The induction step is given by the second axiom. If we assume that x is in the domain of p, that x satisfies p and that x is in the domain of the composed function $g \circ while: M \multimap M$, i.e. that x is in the domain of g and $g(x)$ is in the domain of 'while', then we are able to conclude from the second axiom that x is in the domain of 'while', too. We can determine the value of $while(x)$ if we assume that $g(x)$ is in the domain of while' because of the first axiom. In that case we would obtain $while(x) = while(g(x)) = f(g(x))$. If we continue in this way we obtain that x is in the domain of 'while' if and only if there is a natural number n such that $p(x), p(g(x)), \ldots, p(g^n(x))$, $f(g^n(x))$ are all defined, and that $p(x) = p(g(x)) = \ldots = p(g^{n-1}(x)) = true$, and $p(g^n(x)) = false$. Then $while(x) = f(g^n(x))$ holds.

The initiality of the enrichment by definition guarantees that the domain of the operator 'while' does not contain additional elements. We hope to improve the understanding by the following discussion of a slightly modified but incorrect specification of the semantics of the while - statement. This specification WHILE* may result from WHILE by a substitution of the equation $while(g(x)) = while(g(x))$ in the premise of the second axiom by the equation $g(x) = g(x)$.

We will now construct the domain of the partial operator $while(M) \multimap M$ but now defined by WHILE*. Since the first axiom is not modified, again an element x is in the domain of the operator 'while' if $p(x)$ is defined and gives the value false, and if additionally $f(x)$ is defined. If x is an element for which $p(x)$ and $g(x)$ are defined and $p(x) = true$, then the modified version of the second axiom implies that the partial operator 'while' is defined for that x and for $g(x)$. Let M be for instance the set of natural numbers, $p(x) = true$ iff $x \neq 0$, $g(x) = x+1$ for each $x \in \mathbb{N}$, and $f: \mathbb{N} \multimap \mathbb{N}$ any partial function defined on zero. For this actual parameter the specification WHILE defines the partial function $w: \mathbb{N} \multimap \mathbb{N}$ with $dom(w) = \{0\}$ and $w(0) = f(0)$. If we would start with $x = 1$, then the first axiom of WHILE is not immediately applicable because of $p(x) = true$. According to the second axiom of WHILE

while(1) would be defined if while(1+1) is defined. If we want to prove that while(2) is defined, we have to prove that while(2+1) is defined and this would go on infinitely. This shows that for no natural number x with $x \neq 0$ there is a proof of the applicability of 'while' to x on the basis of the given axioms of WHILE . But if we are looking for a proof of the applicability of 'while' to x with $x \neq 0$ on the basis of the axioms of $WHILE^*$, the situation is completely different.

Since g(1) is defined we would now obtain that **both** while(1) and while(g(1)) are defined and that both values are equal, i.e. while(1) = while(2). In the same way we would obtain for any $x \neq 0$ that while(x) and while(g(x)) = while(x+1) are defined and equal. Evidently there are many total functions while: $\mathbb{N} \longrightarrow \mathbb{N}$ with while(0) = f(0) and while(1) = while(2) = ... = while(x) = ... Since there is no axiom that implies the equality of while(1) with any natural number, the initiality of the enrichment by definition in $WHILE^*$ causes the creation of a new object of sort M . The creation of a new object contradicts the persistency so that not each actual parameter of $WHILE^*$ can be extended to a model of $WHILE^*$.

We hope that the similarity of the axioms in WHILE with the ordinary explanation of the semantics of a while-statement has been observed. We interpret this similarity as a justification of the introduced calculus of functional enrichments of PADT's.

Before we will deal with the actualization of formal parameters of an initial canon we specify a further example for which the class of actual parameters is of greater interest than in the previous specifications.

The object of the specification is some kind of an inhomogeneous storage. For this inhomogeneous storage besides the sets of addresses and values a set of types and some additional functions may be given that associate with each address and with each value an object of sort 'Type'. In addition the equality of addresses may be given. This system of sets and functions is considered as a parameter of the inhomogeneous storage. On the basis of that parameter we define the set of all states of the storage and we define store- , read- , and remove-operations. For these operations we assume that a value x can be stored by an address a only if a and x are both of the same type. Evidently, one can only read a value by an address, if before under that address any value has been stored. These requirements imply that both operations become partial ones. The remove-operation will be specified as a total operation that will not change the state of the storage if we try to remove an information by an address under which before no value has been stored.

```
STORAGE is BOOL with requirement
       sorts  Address, Value, Type
       oprn   a-type(Address)  → Type
              v-type (Value)  → Type
              equal(Address,Address)  → Bool
       axioms  a,a1,a2: Address
       equal(a,a) = true
       if  equal(a1,a2) = true  then  a1 = a2
       with definition
       sorts   Storage
       oprn  initiate  → Storage
             written(Address,Storage)  → Bool
             store(Storage,a:Address,x:Value
                   iff  a-type(a) = v-type(x) )  → Storage
             read(a:Address,s:S+orage
                   iff  written(a,s) = true )  → Value
             remove(Address,Storage)  → Storage
       axioms  a,a1,a2: Address,  s: Storage,  x,y: Value
       written(a,initiate) = false
       if written(a1,s) = false, a-type(a2) = v-type(x)  then
                   written(a1,store(s,a2,x)) = equal(a1,a2)  fi
       if a-type(a)=v-type(x), a-type(a)=v-type(y)  then
                   store(store(s,a,x),a,y) = store(s,a,y)  fi
       if a-type(a1)=v-type(x), a-type(a2)=v-type(y)  then
                   store(store(s,a1,x),a2,y)=store(store(s,a2,y),a1,x)
       if a-type(a) = v-type(x) then
                   remove(a,store(s,a,x)) = s  fi
       if written(a,s) = false then  remove(a,s) = s
       if a-type(a)=v-type(x)  then  read(store(s,a,x)) = x
   end STORAGE
```

In this specification explicit domain conditions within the declaration of the partial operators are used as distinct from the initial canon WHILE . Therefore we payed attention that no additional domain conditions are caused by the axioms.

By means of the actualization of the parameter part one can derive different kinds of storages. It is important that actual parameters themselves may be PADT's. This possibility enables stepwise refinements of PADT's .

In the following the abstract storage will be actualized by setting the disjoint union of the sets of integers, truth values, and strings over an arbitrary alphabet for the parameter sort 'Value'. The parameter sort 'Address' will be actualized by a set of strings overs a possibly different alphabet, and we will finally set the two-element set {static,dynamic} for the parameter sort 'Type'. Therefore the attribute functions 'a-type' and 'v-type' associate with each address and each value one of these both attributes. This actualization uses another parameterized abstract data type as actual parameter so that by further actualizations both the set of characters of the strings that form the addresses and the set of characters that form a part of the values may by actualized. To achieve the sketched actualization we specify at first the actual parameter.

```
PARAMETER1  is  NAT  with requirement
        sorts  Character
        oprn   c-eq(Character,Character)  ——▸  Bool
        axioms  x,y: Character
        c-eq(x,x) = true
        if  c-eq(x,y)=true  then  x=y
        with definition
        sorts  String
        oprn   nil  ——▸  String
               append(String,Character)  ——▸  String
               s-eq(String,String)  ——▸  Bool
        axioms  x,y: Character,  u,v: String
        s-eq(u,u) = true
        s-eq(u,v) = s-eq(v,u)
        s-eq(u,append(u,x)) = false
        s-eq(append(u,x),append(v,y)) = (s-eq(u,v)   c-eq(x,y))
        with definition
        sorts  Union, Kind
        oprn   static, dynamic  ——▸  Kind
               b-inj(Bool)  ——▸  Union
               n-inj(Nat)  ——▸  Union
               s-inj(String)  ——▸  Union
               u-attr(Union)  ——▸  Kind
        axioms  x: Bool, y: Nat, v: String
        u-attr(b-inj(x)) = static
        u-attr(n-inj(y)) = static
        u-attr(s-inj(v)) = dynamic
  end PARAMETER1
```

For an actualization of the parameter sort 'Character' in the initial canon PARAMETER1 we define a special set of characters by the initial canon PARAMETER2.

PARAMETER2 <u>is</u> BOOL <u>with definition</u>
<u>sorts</u> A
<u>oprn</u> a,b,c $\longrightarrow$ A
2-eq(A,A) $\longrightarrow$ Bool
<u>axioms</u> x: A
2-eq(x,x) = true
2-eq(a,b) = 2-eq(a,c) = 2-eq(b,a) = 2-eq(b,c) = 2-eq(c,a) =
= 2-eq(c,b) = false
<u>end</u> PARAMETER2

After the specification of a PADT and an ADT that will be used as actual parameters afterwards we introduce the last operation for the construction of an initial canon. This construction in called 'Modification'. The general syntactic scheme of a modification is the following:

$\mathbb{C}$ <u>is</u> $\mathbb{C}_1$ <u>modified</u> [<u>by</u> $\mathbb{C}_2$] <u>according</u>
S_1 <u>for</u> S_1^*
. . . .
(+) S_n <u>for</u> S_n^*
σ_1 <u>for</u> σ_n^*
. . . .
σ_m <u>for</u> σ_m^*
<u>end</u> $\mathbb{C}$

such that:

(i) $S_1^*,\ldots,S_n^*$ are sort names of the initial canon $\mathbb{C}_1$;

(ii) $\sigma_1^*,\ldots,\sigma_m^*$ are operators of the initial canon $\mathbb{C}_1$;

(iii) S_i and σ_j for $i \in \{1,\ldots,n\}$, $j \in \{1,\ldots m\}$ are either sort names and operators of $\mathbb{C}_2$ or they are new names of sorts and operators different from all sort names and operators of $\mathbb{C}_1$ and $\mathbb{C}_2$ respectively;

(iv) The modification list, i.e. the part between <u>according</u> and <u>end</u>, is compatible with the operator declarations in $\mathbb{C}_1$ and $\mathbb{C}_2$, i.e. if $\sigma_i^*(S_{i_1}^*,\ldots,S_{i_k}^*) \longrightarrow S_{i_0}^*$ is an operator declaration in $\mathbb{C}_2$ then the associated operator in $\mathbb{C}_1$ is declared $\sigma_i(S_{i_1},\dots,S_{i_k}) \longrightarrow S_{i_o}$. The compatibility of the modification list with the operator declarations in $\mathbb{C}_1$ and $\mathbb{C}_2$ includes the requirement that for every operator $\sigma_i^*(S_{i_1}^*,\dots,S_{i_k}^*) \longrightarrow S_{i_o}^*$ contained in the right-hand side of the modification list each sort name $S_{i_j}^*$, $j=0,1,2,\dots,k$, is either a common sort name of $\mathbb{C}_1$ and $\mathbb{C}_2$ or it is contained in the right-hand side of the modification list.

(v) The modification list preserves enrichments by definitions of the initial canon $\mathbb{C}_1$.

The initial canon $\mathbb{C}_1$ is called the modified canon of the modification (+), and the initial canon $\mathbb{C}_2$ is called the modifying canon. As indicated by rectangular brackets the part '<u>by</u> $\mathbb{C}_2$' of a modification (+) may be omitted. In this case the left-hand side of the modification list can contain only new names of sorts and new operators, i.e. the modification is reduced to a simple renaming of some sort names and some operators of $\mathbb{C}_1$. Formally the modification list must define a canon morphism (see Section 4.3) from a subcanon $\mathbb{C}_o$ of $\mathbb{C}_1$ into $\mathbb{C}_2$. The subcanon $\mathbb{C}_o$ is generated by the sort names and operators of the right-hand side of the modification list.

The initial canon $\mathbb{C}$ resulting from a modification (+) of $\mathbb{C}_1$ by $\mathbb{C}_2$ according to a given modification list is constructed in the following way: The subcanon $\mathbb{C}_o$ of $\mathbb{C}_1$, generated by $S_1^*,\dots,S_n^*$ and $\sigma_1^*,\dots,\sigma_m^*$, will be substituted by the image $\mu(\mathbb{C}_o)$ of $\mathbb{C}_o$ in $\mathbb{C}_2$, where μ denotes the canon morphism $\mu\colon \mathbb{C}_o \longrightarrow \mathbb{C}_2$ defined by the modification list. This substitution transforms the initial canon $\mathbb{C}_1$ into the initial canon $\mathbb{C}_1^*$. In the next step the initial canons $\mathbb{C}_1^*$ and $\mathbb{C}_2$ will be joint. Note that equal sort names and equal operators of $\mathbb{C}_1^*$ and $\mathbb{C}_2$ will not be duplicated. This requires a careful use of the modification to avoid conflicts.

Since every enrichment by definition within $\mathbb{C}_o$ will be transformed by $\mu\colon \mathbb{C}_o \longrightarrow \mathbb{C}_2$ into a corresponding enrichment by definition within $\mu(\mathbb{C}_o) \subseteq \mathbb{C}_2 \subseteq \mathbb{C}$, it is impossible to change an enrichment by definition into an enrichment by requirement by means of a modification. It is just the other way round the aim of modifications to transform enrichments by requirements within $\mathbb{C}_1$ into enrichments by definition within the resulting initial canon $\mathbb{C}$. This implies that each modification of an initial canon that does not contain any enrichment by requirement can be only a renaming of a subcanon of $\mathbb{C}_1$ followed by a union with $\mathbb{C}_2$. However, this situation should be made more explicit by a construction like

$\mathbb{C}$ <u>is</u> $\mathbb{C}_1$ <u>modified according</u> ⟨modification list⟩ <u>with</u> $\mathbb{C}_2$.

The modification

```
PARAMETER3  is  PARAMETER1  modified by  PARAMETER2  according
            A  for  Character
            2-eq  for  e-eq
  end PARAMETER3
```

transforms the PADT given by the specification PARAMETER1 into the specification PARAMETER3 of an ADT, since the enrichment by requirement in PARAMETER1 is transformed into an enrichment by definition. We will demonstrate the effect of a modification by this example. The subcanon $\mathbb{C}_o$ of PARAMETER1, generated by A and 2-eq is given by:

```
Co  is  BOOL  with requirement
        sorts  Character
        oprn  c-eq(Character,Character)  ——→  Bool
        axioms  x,y: Character
        c-eq(x,x) = true
        if  c-eq(x,y)=true  then  x=y
  end Co
```

Even if the modification list does not contain the sort name 'Bool' this sort name is contained in $\mathbb{C}_o$ because of the occurrence of the operator 'c-eq' in the modification list. The image $\mu(\mathbb{C}_o)$ in the specification PARAMETER2 is the initial canon PARAMETER2 itself. This implies that $\mathbb{C}_1^*$ becomes $\mathbb{C}$, so that we obtain the following specification as the result of the modification:

```
PARAMETER3  is  NAT  with definition
        sorts  A
        oprn  a,b,c  ——→  A
              2-eq(A,A)  ——→  Bool
        axioms  x: A
        2-eq(x,x) = true
        2-eq(a,b) = 2-eq(a,c) = 2-eq(b,a) = 2-eq(b,c) = 2-eq(c,a) =
                  = 2-eq(c,b) = false
        with definition
        sorts  String
```

```
oprn  nil  ——▶ String
      append(String,A)  ——▶  String
      s-eq(String,String)  ——▶  Bool
axioms  x,y: A,  u,v: String
   . . . . . . . . . . . . . .
the same continuation as in PARAMETER1
   . . . . . . . . . . . . . .
end PARAMETER3
```

We obtain a prototype of the ADT specified by PARAMETER3 if we take for 'Nat' the set of natural numbers, for 'Bool' the set of truth values, for 'A' the three-element alphabet $\{a,b,c\}$, for 'String' the set $\{a,b,c\}^*$ of all finite sequences of elements of $\{a,b,c\}$, for 'Union' the set $\mathbb{N} \cup \{true, false\} \cup \{a,b,c\}^*$, and for 'Kind' the two-element set $\{static, dynamic\}$. The semantic interpretation of the operators can easily be deduced from the names of the operators.

The following modification of STORAGE is a first step in the direction to the intended actualization.

```
STORAGE1  is  STORAGE  modified by  PARAMETER1  according
          Union  for  Value
          Kind   for  Type
          u-attr for  v-type   end  STORAGE1
```

An expanded version of the initial canon STORAGE1 is given by the following specification.

```
STORAGE1  is  PARAMETER1  with  requirement
     sorts  Address
     oprn  a-type(Address)  ——▶  Kind
           equal(Address,Address)  ——▶  Bool
     axioms  a,a1,a2: Address
     equal(a,a) = true
     if  equal(a1,a2) = true  then  a1 = a2
     with definition
        . . . . . . . . . . . . . .
     the same continuation as in  STORAGE
        . . . . . . . . . . . . . .
  end  STORAGE1
```

In this specification the sort names 'Character' and 'Address' represent parameters. Both sort names may be interpreted by arbitrary sets. The intended actualization of STORAGE will be obtained by a modification of STORAGE1 using the following modifying canon PARAMETER4.

PARAMETER4 <u>is</u> <u>requirement</u>
        <u>sorts</u> Sign
        <u>with</u> <u>definition</u>
        <u>sorts</u> Sequence
        <u>oprn</u> nil ⟶ Sequence
             add(Sequence,Sign) ⟶ Sequence
  <u>and</u> PARAMETER4

The described actualization of STORAGE is then given by

STORAGE2 <u>is</u> STORAGE1 <u>modified</u> <u>by</u> PARAMETER4 <u>according</u>
        Sequence <u>for</u> Address
  <u>end</u> STORAGE2

The following modification of STORAGE yields a more restricted actualization as the previous one by restricting addresses to natural numbers.

STORAGE3 <u>is</u> STORAGE <u>modified</u> <u>by</u> PARAMETER1 <u>according</u>
        Union <u>for</u> Value
        Kind <u>for</u> Type
        u-attr <u>for</u> v-type
        Nat <u>for</u> Address
        eq <u>for</u> equal
  <u>end</u> STORAGE3

However, STORAGE3 does not specify an ADT since the sort name 'Character' and the operator 'a-type(Nat) ⟶ Kind' represent parameters.

With these examples we conclude Section 1.2 which is aimed as informal representation of the different means of writing specifications that may contain partial operations.

1.3. The Definition of the Semantics of a Functional Language over arbitrary PADT's as a Complex Example

The main point of the following section is the specification of a complex example. But the search for a suitable example is not a simple task. Examples for which there are mathematical descriptions do not require formal specifications. Examples for which there are no adequate mathematical descriptions are hard to explain independently of a formal specification. Therefore we will use an example that can roughly be presented in an informal way, but a detailed presentation will require formal specifications.

We start with the informal presentation of the functional language over any PADT. The object of that language is the construction of tuple-valued partial functions

$$t: S_1S_2...S_n \longrightarrow\!\!\!\!\!\!\circ\;\; R_1R_2...R_m \quad , n,m \geq 0 \; ,$$

which will be called 'tuple-functions' for short, from fundamental operators of a fixed but arbitrary initial canon $\mathbb{B}$. The semantics of such a tuple-function over $\mathbb{B}$ depends on the interpretation of the sort names and of the operators of $\mathbb{B}$. Hence, each tuple-function represents a parameterized algorithm that produces an m-tuple of values from an n-tuple of arguments, where the i-th component of the result is of sort R_i . However, the semantics of the single constructions of tuple-functions over $\mathbb{B}$ has to be independent of the interpretation of $\mathbb{B}$. The functional language is basically given by the following constructions of tuple-functions:

- composition,
- looping,
- branching,
- source tupling,
- projection.

The composition of tuple-functions is a partial composition, since the tuple-functions $t_1: u_1 \longrightarrow\!\!\!\!\!\!\circ\;\; v_1$, $t_2: u_2 \longrightarrow\!\!\!\!\!\!\circ\;\; v_2$, where u_i, v_i for $i=1,2$ denote finite sequences of sort names of $\mathbb{B}$, can be composed if and only if $v_1 = u_2$. The result of the composition is then a tuple-function $t_1;t_2: u_1 \longrightarrow\!\!\!\!\!\!\circ\;\; v_2$.

The looping construction requires three arguments, $p: u \longrightarrow\!\!\!\!\!\!\circ\;\; Bool$, $f: u \longrightarrow\!\!\!\!\!\!\circ\;\; u$, $g: u \longrightarrow\!\!\!\!\!\!\circ\;\; u$ and corresponds to the well-known construction

'while p do f else g': $u \longrightarrow\!\!\!\!\!\!\circ\;\; u$.

The arguments of the looping construction have to satisfy the following conditions. The input tuples of all three arguments are of equal type, given by u, the output sequence of p is the sequence of length one given by 'Bool', and the output sequences of f and g are equal to the common input sequence of p,f,g.

The branching construction corresponds to another well-known construction of programming languages, namely to

$$\text{'if } p \text{ then } g \text{ else } f\text{'}: u \multimap v$$

where $p: u \multimap \text{Bool}$, $g: u \multimap v$, $f: u \multimap v$ are the arguments. Again it is required that p,f,g have equal input sequences, and that the sequence of length one 'Bool' is the output sequence of p. But now we require only that f and g have equal output sequences which may be different from the common input sequence of p, f, g.

The source tupling is applicable to any finite sequence of tuple-functions

$$t_1: u \multimap v_1,\ t_2: u \multimap v_2,\ \ldots,\ t_n: u \multimap v_n$$

with a common input sequence. The result is the tuple function

$$(t_1,\ldots,t_n): u \multimap v_1v_2\ldots v_n$$

where the output sequence $v_1v_2\ldots v_n$ of the result is composed from the output sequences $v_1,v_2,\ldots,v_n$ by concatenation. Semantically

$$(t_1,\ldots,t_n) : u \multimap v_1v_2\ldots v_n$$

corresponds to the parallel execution of $t_1,t_2,\ldots,t_n$ that comes to a result if each of the arguments comes to a result. Therefore the domain of $(t_1,\ldots,t_n)$ is the intersection of the domains of the arguments. The source tupling of the empty sequence of tuple-functions with common input sequence u will be denoted by

$$(): u \multimap \lambda$$

where λ denotes the empty sequence of sort names of $\mathbb{B}$. Semantically $(): u \multimap \lambda$ corresponds to the forgetting of each argument.

Finally, for each sequence of sort names of $\mathbb{B}$,

$$u: \{1,2,\ldots,n\} \longrightarrow \text{sorts}\,\mathbb{B}$$

and each natural number $i \in \{1,\ldots,n\}$ there is a so-called i-th projection of u with

$$p_i: u \multimap u(i),$$

i.e. with the input sequence u and with the sequence of length one $u(i)$ as output sequence. The i-th projection of u singles out the i-th argument and it is therefore a total function for each u and each natural number i with $i \leq \text{length}(u)$.

We demonstrate the expressiveness of the sketched functional language over PADT's by some examples. For these examples we set $\mathbb{B}$ = NAT. If $r: \mathbb{N} \times \mathbb{N} \longrightarrow \mathbb{N}$ denotes the binary total function with

$$x = y \cdot q + r(x,y) \quad , 0 \leq r(x,y) \leq y, \text{ and } \quad y \neq r(x,y),$$

and if $p_1: \mathbb{N} \times \mathbb{N} \longrightarrow \mathbb{N}$, $p_2: \mathbb{N} \times \mathbb{N} \longrightarrow \mathbb{N}$ denote the projections to the first and second argument respectively, then the Euclidian algorithm can be expressed by

(*) while p_2=0 do (p_2, r) else (p_1, p_2) ; p_1 .

Here p_2=0 stands for

$$p_2 \; ; \text{ nonzero } : \mathbb{N} \times \mathbb{N} \longrightarrow \text{Bool}$$

where nonzero: $\mathbb{N} \longrightarrow$ Bool is the unary predicate with nonzero(x) = = true if and only if $x \neq 0$. Remember that the semicolon denotes the composition of tuple-functions.

Starting with the argument tuple (15,9) the tuple function (*) over NAT produces the following sequence of intermediate results:

$$(15,9) \longmapsto (9,6) \longmapsto (6,3) \longmapsto (3,0) \longmapsto 3 \; .$$

In the preceding looping construction the else-part is of no interest, because the source tupling of the first and second projections forms the identity of pairs of natural numbers. Usually one allows to omit the else-part of a loop expression if its arguments represent the identity. With this convention the factorial function n! can be expressed as follows:

$$(id_{\mathbb{N}}, id_{\mathbb{N}}) \; ; \text{ while } p_2{=}0 \text{ do } (\text{times}, p_2;\text{pred}) \; ; \; p_2$$

where pred: $\mathbb{N} \multimap \mathbb{N}$ denotes the unary partial predecessor function and times: $\mathbb{N} \times \mathbb{N} \longrightarrow \mathbb{N}$ denotes the multiplication.

Even if we used some notations for the preceding examples of tuple-functions over NAT we will not define any specific syntax for the functional language but we will define an abstract syntax that contains the context conditions of the single constructions. The following specification SIGNATURE describes the knowledge on the base canon that is used within the abstract syntax.

SIGNATURE <u>is</u> BOOL <u>with</u> <u>requirement</u>
    <u>sorts</u> Sorts, Operators
    <u>oprn</u> bool $\longrightarrow$ Sorts
    <u>with</u> <u>definition</u>
    <u>sorts</u> Sorts^*
    <u>oprn</u> nil $\longrightarrow$ Sorts^*
        app(Sorts^*,Sorts) $\longrightarrow$ Sorts^*

```
    with requirement
    oprn  source(Operators) ——> Sorts*
          target(Operators) ——> Sorts
end SIGNATURE
```

The following enrichment by definition of SIGNATURE represents the parameterized abstract syntax of the functional language.

```
FUNCTIONS is SIGNATURE with NAT with definition
    oprn  lenght(Sorts*) ——> Nat
          concat(Sorts*,Sorts*) ——> Sorts*
          val(w:Sorts*,i:Nat iff i≤length(w)) ——> Sorts
    axioms  s: Sorts,  w,w1,w2: Sorts*
    length(nil) = zero
    length(app(w,s)) = succ(length(w))
    concat(w,nil) = w
    concat(w1,app(w2,s)) = app(concat(w1,w2),s)
    if  w = concat(app(w1,s),w2), length(app(w1,s))=i  then
                        val(w,i) = s  fi
    with definition
    sorts  F, F*
    oprn  in(F) ——> Sorts*
          out(F) ——> Sorts*
          base(Operators) ——> F
          compose(f:F, g:F, iff out(f)=in(g)) ——> F
          prj(w:Sorts*, i:Nat iff i ≤ length(w)) ——> F
          loop(p:F, g:F, f:F iff  in(p)=in(g)=in(f)=out(f)=
               =out(g), out(p)=app(nil,bool)) ——> F
          branch(p:F, g:F, f:F iff  in(p)=in(g)=in(f),
                 out(f)=out(g),out(p)=app(nil,bool)) ——> F
          away(Sorts*) ——> F
          first(F) ——> F*
          add(f*:F*, f:F iff in*(f*)=in(f)) ——> F*
          in*(F*) ——> Sorts*
          tuple(F*) ——> F
    axioms  x: Operators, p,f,g,h: F, g*,f*: F*, i: Nat, w: Sorts*
    in(base(x)) = source(x)
    out(base(x)) = app(nil,target(x))
    if  h = compose(f,g)  then  in(h) = in(f)
    if  h = compose(f,g)  then  out(h) = out(g)
    if  f = prj(w,i)  then  in(f) = w
    if  f = prj(w,i)  then  out(f) = app(nil,val(w,i))
    if  h = loop(p,f,g) then  in(h) = in(f)
```

```
if  h = loop(p,f,g)  then  out(h) = in(h)
if  h = branch(p,f,g)  then  out(h) = out(f)
if  h = branch(p,f,g)  then  in(h) = in(f)
in(away(w)) = w
out(away(w)) = nil
in*(first(f)) = in(f)
if  g* = add(f*,f)  then  in*(g*) = in*(f*)
in(tuple(f*)) = in*(f*)
out(tuple(first(f))) = out(f)
if g* = add(f*,f)  then
      out(tuple(g*)) = concat(out(tuple(f*),out(f))  fi
end  FUNCTIONS
```

This specification needs some explanations. In the first enrichment by definition some auxiliary operations on sequences of elements of 'Sorts' are specified. The operator 'length($Sorts^*$) $\longrightarrow$ Nat' gives the length for each $w \in Sorts^*$, and 'concat($Sorts^*$,$Sorts^*$) $\longrightarrow$ $Sorts^*$' represents the concatenation of sequences of sorts, and 'val($Sorts^*$,Nat) $-\!\circ\!\!\rightarrow$ Sorts yields the i-th element of a sequence. The second enrichment by definition also contains some auxiliary operations and the auxiliary sort F^* which represents finite non-empty sequences of tuple-functions such that all tuple-functions in a sequence have equal inputs whereas tuple-functions contained in different sequences may have different inputs. The operators 'in(F) $\longrightarrow$ $Sorts^*$' and 'out(F) $\longrightarrow$ $Sorts^*$' associate with each tuple-function its input and output sequence of sorts. These operators are necessary for the context conditions of the single constructions of tuple-functions. These constructions are represented by the operators 'base','compose','prj', 'loop','branch','away', and 'tuple'.

Besides the constructions mentioned in the informal presentation the formal specification contains the construction 'base(Operator) $\longrightarrow$ F' that transforms each operator into a tuple-function. The operator 'away($Sorts^*$) $\longrightarrow$ F' represents the source tupling of empty sequences of tuple-functions, and the operator 'tuple(F^*) $\longrightarrow$ F' represents the source tupling of non-empty sequences of tuple-functions.

The semantics of the functional language will be defined by means of so-called 'pointed enrichments by defintions of the base canon $\mathbb{B}$'. A pointed enrichment by definition of $\mathbb{B}$ is an enrichment by definition $\mathbb{B}$ together with a distinguished n-tuple of terms over $\mathbb{B}$

$$(t_1 : w -\!\circ\!\!\rightarrow s_1,\ t_2 : w -\!\circ\!\!\rightarrow s_2,\ \dots\ ,\ t_n : w -\!\circ\!\!\rightarrow s_n)$$

such that $s_1, s_2, \dots, s_n \in$ sorts $\mathbb{B}$, and $w \in (\text{sorts } \mathbb{B})^*$ is a sequence of sort names of $\mathbb{B}$. Notice that all terms t_i, $i=1,2,\ldots,n$, of a distinguished n-tuple of terms have $w \in (\text{sorts } \mathbb{B})^*$ as common input, and that each t_i produces values of a sort S_i of the base canon $\mathbb{B}$.

This notion of a distinguished n-tuple of terms over the base canon $\mathbb{B}$ suffices for the definition of the semantics of the constructions 'base', 'compose', 'prj', 'away', and 'tuple'. But this notion is not strong enough for the definition of the semantics of the constructions 'loop' and 'branch'. These constructions are the only two that require really enrichments by definitions. Even if the corresponding enrichments by definitions do not allow the definition of new data types, this incapability is not a serious fault of the functional language. Because of the independece of the semantics of the functional language of the underlying base canon $\mathbb{B}$, the definition of a new data type can be realized by the use of a corresponding enrichment by definition of $\mathbb{B}$ as new base canon.

Let $\mathbb{B}$ be any given base canon. The signature of $\mathbb{B}$ represents a correct interpretation of the initial canon SIGNATURE given by the sets sorts($\mathbb{B}$), operators($\mathbb{B}$) and by the fundamental operations

$$\text{source}^{\mathbb{B}}: \text{operators}(\mathbb{B}) \longrightarrow (\text{sorts}(\mathbb{B}))^*$$

$$\text{target}^{\mathbb{B}}: \text{operators}(\mathbb{B}) \longrightarrow \text{sorts}(\mathbb{B}).$$

The principle of the following semantics definition is based on the universal property of interpretations that satisfy initially a specification. This universal property is studied in detail in Section 3.2. Roughly speaking, one can say: if I is an interpretation that initially satisfies a specification, and if A is any interpretation then the universal property of I implies the existence of exactly one homomorphism $f: I \longrightarrow A$. By means of a homomorphism defined in this way we will define the semantics of the functional language. Any signature Σ, i.e. any interpretation of SIGNATURE, can be extended to an interpretation of FUNCTIONS unique up to isomorphism. For any such interpretation $I(\Sigma)$ that initially satisfies the axioms of FUNCTIONS and any other interpretation A of FUNCTIONS that satisfies the axioms not necessarily in initial way there is precisely one homomorphism $f: I(\Sigma) \longrightarrow A$. Hence, the essence of the semantics definition is the construction of a suitable algebra A such that for each abstract tuple-function $\tau \in I(\Sigma)_F$ the element $f(\tau) \in A_F$ represents the meaning of τ. We will construct the algebra A in such a way that the meaning $f(\tau)$ becomes an initial specification of a functional extension of the base canon $\mathbb{B}$.

The following denotation reduces the denotational amount of dealing with n-tuples of terms and with sequences of n-tuples of terms on enrichments $\hat{\mathbb{B}}$ of $\mathbb{B}$. If $\hat{\mathbb{B}}$ is any enrichment by definition of the given base canon $\mathbb{B}$, and if $(t_1 : w \multimap S_1, \ldots, t_n : w \multimap S_n)$ is an n-tuple of terms over the signature of $\hat{\mathbb{B}}$ with $w \in \text{sorts}(\mathbb{B})^*$ and $S_1, \ldots, S_n \in \text{sorts}(\mathbb{B})$, then this n-tuple will be denoted by

$$\underline{t} : w \multimap u \quad \text{with} \quad u = S_1 \ldots S_n \in \text{sorts}(\mathbb{B})^*$$

and it will be called a 'w-term tuple' on $\hat{\mathbb{B}}$. The prefix w indicates the common input domain w of all terms of the w-term tuple. By

$$\underline{t}(i) : w \multimap u(i) \quad \text{for each} \quad 1 \leq i \leq \text{length}(u) = n$$

we denote the i-th component of $\underline{t} : w \multimap u$.

For the definition of the set A_{F^*} we need sequences of w-term tuples over enrichments by definition $\hat{\mathbb{B}}$ of $\mathbb{B}$. For that reason we use

$$\underline{t}^{\#} : w \multimap u_1 \# u_2 \# \ldots \# u_r \quad , r \geq 0,$$

as denotation of the sequence of w-term tuples

$$(\underline{t}^{\#}(1) : w \multimap u_1, \ \underline{t}^{\#}(2) : w \multimap u_2, \ldots, \ \underline{t}^{\#}(r) : w \multimap u_r) .$$

$\underline{t}^{\#} : w \multimap u_1 \# u_2 \# \ldots \# u_r$ is called a 'w-term tuple sequence', and the natural number r is called the length of the w-term tuple sequence. Both notations can be combined so that

$$(\underline{t}^{\#}(i))(j) : w \multimap u_i(j)$$

denotes the j-th component of the i-th element of the w-term tuple sequence $\underline{t}^{\#} : w \multimap u_1 \# u_2 \# \ldots \# u_r$.

Now we are able to define the algebra A that defines the semantics of the functional language.

First of all we set

$$A_{\text{Sorts}} = \text{sorts}(\mathbb{B}) , \quad A_{\text{Operators}} = \text{operators}(\mathbb{B}),$$

$$\text{source}^A = \text{source}^{\mathbb{B}} , \quad \text{target}^A = \text{target}^{\mathbb{B}},$$

$$A_{\text{Nat}} = \mathbb{N}, \quad A_{\text{Bool}} = \{\text{true}, \text{false}\} , \quad A_{\text{Sorts}^*} = \text{sorts}(\mathbb{B})^*,$$

and nil^A, app^A, length^A, concat^A, and val^A are interpreted as the empty sequence of sort names of $\mathbb{B}$, as the appending of a sort name to a given sequence of sort names, as the length of sequences of sort names, as the concatenation of sequences of sort names, and as the selection of the i-th component of a sequence, respectively.

The interpretation of the remaining sorts F and F^* and of the remaining operators is the kernel of the definition of semantics.

We start with the definitions of the sets A_F and A_{F^*} :

$$A_F = \{ (\hat{\mathbb{B}}, \underline{t}: w \longrightarrow u) \mid \text{such that } \hat{\mathbb{B}} \text{ is an enrichment by definition of } \mathbb{B} \text{, and } \underline{t}: w \longrightarrow u \text{ with } w,u \in sorts(\mathbb{B})^* \text{ is a w-term tuple on } \hat{\mathbb{B}} \}$$

$$A_{F^*} = \{ (\hat{\mathbb{B}}, \underline{t}^{\#} : w \longrightarrow u_1\#u_2\#\ldots\#u_r) \mid \text{such that } \hat{\mathbb{B}} \text{ is an enrichment by definition of } \mathbb{B} \text{, and } \underline{t}^{\#}: w \longrightarrow u_1\#\ldots\#u_r \text{ is a w-term tuple sequence on } \hat{\mathbb{B}} \text{ with } r \in \mathbb{N} \text{, and } w,u_1,\ldots,u_r \in sorts(\mathbb{B}) \} .$$

The interpretations of the operators 'in','out','in^*' are naturally defined by

$$in^A(\hat{\mathbb{B}}, \underline{t}: w \longrightarrow u) = w ,$$
$$out^A(\hat{\mathbb{B}}, \underline{t}: w \longrightarrow u) = u ,$$
$$(in^*)^A(\hat{\mathbb{B}}, \underline{t}^{\#}: w \longrightarrow u_1\#u_2\#\ldots\#u_r) = w .$$

Since the sort F^* represents only an auxiliary construction of special sequences of w-term tuples the interpretation of the operators 'first' and 'add' is naturally given by

$$first^A(\hat{\mathbb{B}}, \underline{t}: w \longrightarrow u) = (\hat{\mathbb{B}}, \underline{t}^{\#}: w \longrightarrow u)$$

where $\underline{t}: w \longrightarrow u$ is interpreted as w-term tuple sequence of length one, and

$$add^A((\hat{\mathbb{B}}_1, \underline{t}_1^{\#}: w \longrightarrow u_1\#\ldots\#u_r),(\hat{\mathbb{B}}_2, \underline{t}: w \longrightarrow u)) =$$
$$= (\hat{\mathbb{B}}_1 \underline{\text{ with }} \hat{\mathbb{B}}_2, \underline{t}_2^{\#}: w \longrightarrow u_1\#\ldots\#u_r\#u)$$

such that $length(\underline{t}_2^{\#}) = length(\underline{t}_1^{\#})+1$, and $\underline{t}_2^{\#}(i) = \underline{t}_1^{\#}(i)$ for $1 \leq i \leq r = length(\underline{t}_1^{\#})$, $\underline{t}_2^{\#}(r+1) = \underline{t}$.

The most interesting part of the definition of the semantics is the interpretation of the operators 'base','compose','prj','loop','branch', 'away', and 'tuple' since these operators represent the constructions of tuple-functions.

Semantics of 'base':

For each operator $\sigma: w \longrightarrow s$ in the signature of the initial canon $\mathbb{B}$ we define

$base^A(\sigma) = (\mathbb{B}, \sigma: w \longrightarrow s)$

where the operator is considered as a term over $\mathbb{B}$, and that term is considered as a w-term tuple of one component.

Semantics of 'compose':

The elements $(\hat{\mathbb{B}}_1, \underline{t}_1: w_1 \longrightarrow u_1)$, $(\hat{\mathbb{B}}_2, \underline{t}_2: w_2 \longrightarrow u_2)$ of A_F are in the domain of the partial composition ,compose^A, if and only if $u_1 = w_2$. In that case we set

$$compose^A((\hat{\mathbb{B}}_1, \underline{t}_1: w_1 \longrightarrow u_1),(\hat{\mathbb{B}}_2, \underline{t}_2: u_1 \longrightarrow u_2)) =$$
$$= (\mathbb{B}_1 \ \underline{with}\ \mathbb{B}_2, \underline{t}: w_1 \longrightarrow u_2)$$

with $\underline{t}(i) = [\underline{t}_2(i)](\underline{t}_1)$ for $1 \leq i \leq length(u_2)$, where

$$[\underline{t}_2(i)](\underline{t}_1): w_1 \longrightarrow u_2(i)$$

denotes the term that results from $\underline{t}_2(i): u_1 \longrightarrow u_2(i)$ by the simultaneous substitution of $\underline{t}_1(j)$ for the j-th variable in $\underline{t}_2(i): u_1 \longrightarrow u_2(i)$ for each j with $1 \leq j \leq length(u_1)$.

We demonstrate this composition of w-term tuples by a simple example Let $\underline{t}_1$ = (g.c.d: Nat X Nat $\longrightarrow$ Nat, plus: Nat X Nat $\longrightarrow$ Nat) and $\underline{t}_2$ = (less: Nat X Nat $\longrightarrow$ Bool, times: Nat X Nat $\longrightarrow$ Nat, l.c.m: Nat X Nat $\longrightarrow$ Nat) be term tuples on NAT with BOOL. The composition of $\underline{t}_1$ with $\underline{t}_2$ yields a term tuple $\underline{t}$: Nat X Nat $\longrightarrow$ Bool X Nat X Nat that produces a triple $(y_1, y_2, y_3) \in \{true, false\} \times \mathbb{N} \times \mathbb{N}$ for each pair $(x,y) \in \mathbb{N} \times \mathbb{N}$ in such a way that

y_1 = g.c.d(x,y) $\leq$ (x+y) (=true for all x,y),
y_2 = g.c.d(x,y) $*$ (x+y),
y_3 = l.c.m(g.c.d(x,y),x+y) .

Semantics of 'prj':

In the specifications of the preceding sections projections that pick out some component of an n-tuple of values do not explicitly occur. Instead of those projections we used variables, and the evaluation of a variable is the equivalent to the application of a projection. However in the examples within the informal presentation of the functional language we used explicit denotations for projections. We illustrate the equivalence between the use of variables and the use of projections by a simple example. Let us consicer a declaration of the form

x,y: Nat, r: String, a,b: Bool

then the variable r represents the third projection

$$p_3: \text{Nat},\text{Nat},\text{String},\text{Bool},\text{Bool} \longrightarrow \text{String}$$

whereas the variable y would represent the projection

$$p_2: \text{Nat},\text{Nat},\text{String},\text{Bool},\text{Bool} \longrightarrow \text{Nat}$$

that picks out the second element of a given 5-tuple.
For the definition of the semantics of the functional language we assume the existence of projections. For each sequence w of sort names of $\mathbf{B}$ and each natural number i with $i \leq \text{length}(w)$

$$p_i^w: w \multimap w(i)$$

denotes the i-th projection. With this convention we can define

$$\text{prj}^A(w,i) = (\mathbf{B}, p_i^w: w \multimap w(i)) .$$

Semantics of 'away':

The intended semantics of this construction was the tuple-function that takes away each argument. In the framework of w-term tuples this construction can be represented by the zero-tuple of w-terms that will be denoted by

$$\emptyset: w \multimap \text{nil}^A$$

where nil^A denotes the empty sequence of sort names. With this notation we can define

$$\text{away}^A(w) = (\mathbf{B}, \emptyset: w \multimap \text{nil}^A) .$$

Semantics of 'tuple':

Up to this point the defined constructions generate only tuple-functions that produce single values or one-tuples. The construction that generates actual tuple-functions is the source tupling represented by the operator 'tuple'. In the framework of w-term tuples this conctruction transforms a sequence of w-term tuples into one w-term tuple. The formal definition is as follows:

$$\text{tuple}^A(\mathbf{B}, \underline{t}^{\#}: w \multimap u_1{=}u_2{=}\ldots{=}u_r) = (\mathbf{B}, \underline{t}: w \multimap u)$$

where $u = u_1u_2\ldots u_r \in \text{sorts}(\mathbf{B})^*$ is the concatenation of $u_1,\ldots,u_r$, and

$$\underline{t}(k) = (\underline{t}^{\#}(i))(j) \quad \text{if} \quad k = \text{length}(u_1)+\ldots+\text{length}(u_{i-1})+j \quad \text{and}$$
$$\text{length}(u_{i-1}) < k \leq \text{length}(u_i) ,$$

$$\underline{t}(k) = (\underline{t}^{\#}(1))(k) \quad \text{if} \quad 1 \leq k \leq \text{length}(u_1) .$$

Semantics of 'loop':

The semantics of the loop-construction will be defined by a suitable modification of the previously defined initial canon WHILE.

Let $(\hat{\mathbb{B}}_1, t_1 : w \multimap Bool), (\hat{\mathbb{B}}_2, \underline{t}_2 : w \multimap w), (\hat{\mathbb{B}}_3, \underline{t}_3 : w \multimap w)$ be given arguments of the partial operation '$loop^A$' contained in the domain of that partial operation. Then the intended semantics requires the substitution of the parameter sort M in the intial canon WHILE by the sequence $w \in sorts(\mathbb{B})^*$, and of the parameter operators $p(M) \multimap Bool$, $f(M) \multimap M$, $g(M) \multimap M$ by $t_1 : w \multimap Bool$, $\underline{t}_2 : w \multimap w$, $\underline{t}_3 : w \multimap w$ respectively.

This kind of modification requires a slight generalization of the modification of initial canons as defined up to now. On the left-hand side of a modification list we permit now sequences of sort names and w-term tuples of the modified canon. With this generalization we are able to define

$loop^A((\hat{\mathbb{B}}_1, t_1 :w \multimap Bool), (\hat{\mathbb{B}}_2, \underline{t}_2 :w \multimap w), (\hat{\mathbb{B}}_3, \underline{t}_3 :w \multimap w)) =$

=(WHILE <u>modified</u> <u>by</u> ($\hat{\mathbb{B}}_1$ with $\hat{\mathbb{B}}_2$ with $\hat{\mathbb{B}}_3$) <u>according</u>
w <u>for</u> M
t_1 <u>for</u> p
$\underline{t}_2$ <u>for</u> f
$\underline{t}_3$ <u>for</u> g <u>end</u> , while: $M \multimap M$)

If the semantics of the loop-construction should be given by the following flowchart, then this change in the semantics of the loop-construction can easily be realized by the use of a corresponding initial canon LOOP defined in the following.

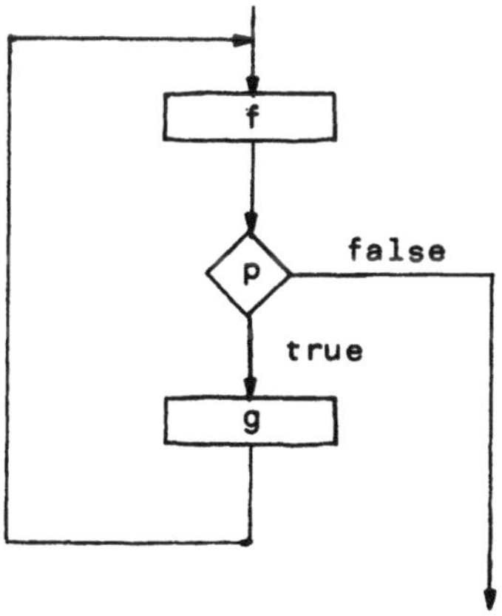

The kernel of the changed definition of the semantics is the specification of the initial canon LOOP.

```
LOOP  is  BOOL  with requirement
        sorts   M
        oprn    p(M) —⊖→ Bool
                f(M) —⊖→ M
                g(M) —⊖→ M
        with definition
        oprn   loop(M) —⊖→ M
        axioms   x: M
        if  p(f(x)) = false then  loop(x) = f(x)
        if  p(f(x)) = true, loop(g(f(x))) = loop(g(f(x)))  then
                      loop(x) = loop(g(f(x)))  fi
  end LOOP
```

Using this specification we define

$$loop^A((\hat{\mathbf{B}}_1, t_1{:}w \multimap Bool),(\hat{\mathbf{B}}_2, \underline{t}_2{:}w \multimap w),(\hat{\mathbf{B}}_3, \underline{t}_3{:}w \multimap w)) =$$

=(LOOP modified by ($\hat{\mathbf{B}}_1$ with $\hat{\mathbf{B}}_2$ with $\hat{\mathbf{B}}_3$) according
w for M
t_1 for p
$\underline{t}_2$ for f
$\underline{t}_3$ for g end , loop: w —⊖→ w)

Semantics of 'branch':

The semantics of the branch-construction will be defined in the same way as the semantics of the loop-construction. Evidently, we have to use another initial canon. We start with the specification of the initial canon BRANCH.

```
BRANCH  is  BOOL  with requirement
        sorts  P,M
        oprn   p(P) —⊖→ Bool
               f(P) —⊖→ M
               g(P) —⊖→ M
        with definition
        oprn  branch(P)  —⊖→ M
        axioms  x: P
        if  p(x) = true, f(x) = f(x) then  branch(x) = f(x)
        if  p(x) = false, g(x) = g(x) then  branch(x) = g(x)
  end  BRANCH
```

Using this specification we define

$$\text{branch}^A((\hat{\mathbb{B}}_1, t_1 : w \rightarrow \text{Bool}), (\hat{\mathbb{B}}_2, \underline{t}_2 : w \rightarrow u), (\hat{\mathbb{B}}_3, \underline{t}_3 : w \rightarrow u)) =$$

$= ($ BRANCH modified by $(\hat{\mathbb{B}}_1$ with $\hat{\mathbb{B}}_2$ with $\hat{\mathbb{B}}_3)$ according

w for P

u for M

t_1 for p

$\underline{t}_2$ for f

$\underline{t}_3$ for g end , branch: P → M)

The preceding semantics of the branching does not assume the applicability of both alternatives. Therefore, a further conceivable semantics of the branching could be given by the following specification BRANCH* defined as follows.

```
BRANCH*  is  BOOL  with requirement
      sorts  P,M
      oprn  p(P) ─→ Bool
            f(P) ─→ M
            g(P) ─→ M
      with definition
      oprn  branch*(P) ─→ M
      axioms  x: P
      if p(x)=false, f(x)=f(x), g(x)=g(x) then  branch*(x)=g(x)
      if p(x)=true, f(x)=f(x), g(x)=g(x) then  branch*(x)=f(x)
  end BRANCH*
```

The initial canon BRANCH* presumes the applicability of both alternatives and of the branching condition as sufficient domain condition of 'branch*(P) → M', and due to the initiality of the last enrichment in BRANCH* this condition is also a necessary domain condition.
•

The homomorphism $f: I(\Sigma) \longrightarrow A$ uniquely defined by the semantic algebra A is not surjective, i.e. not any $(\mathbb{B}, \underline{t}: w \rightarrow u)$ is the meaning of an abstract tuple-function.

Finally we repeat that it is not the aim of this section to introduce a functional language and to contribute to the research work on functional programming, rather than to demonstrate the applicability of structural induction even if the generating constructions are not defined for all arguments, and to use for this demonstration a non-trivial example.

CHAPTER 2

EQUATIONAL PARTIALITY

Whereas in the first chapter it is shown that there are strong motivations for the development of a theory of equationally partial many-sorted algebras, the following chapter develops the algebraic basis of the preceding use of structural induction on partial algebras. According to that aim the following two chapters are a contribution to the universal algebra of partial algebraic structures.

The existence of different sorts of elements in all examples of Chapter one is algebraically reflected by the use of many-sorted algebras (or heterogeneous algebras as they are called by Birkhoff and Lipson (1970)). However, the use of many-sorted algebras causes some notational difficulties. These notational problems could be considerably reduced if we were to use the mathematical language of category theory, but this language is a very abstract one and the uses of abstraction to achieve generality are not the same as in other branches of mathematics and computer science. The theory of categories is a very useful and powerful tool in the hands of a well-trained mathematician. Because most mathematicians and specialists in computer science are not familiar with the basic concepts of category theory, we will not use this tool here. Readers who are interested in category theory should consult the famous book ' Categories for the working mathematician' by MacLane (1971) or a recent paper of Goguen and Burstall (1983). In the following we will use the traditional language of sets, functions, operations, relations, and so on. We deal mostly with S-index families and functions between S-indexed families of sets, where S represents the set of different sorts of objects.

To overcome the notational problems of many-sorted algebras we introduce in the first section the underlying set-theoretical notions and concepts. In the second section we introduce the concepts of total many-sorted algebras, of partial many-sorted algebras, of homomorphisms and of so-called existence equations. The third section is devoted to defining the concept of equational partiality and equationally partial algebras. The fourth section deals with hierarchy conditions for both systems of domain conditions of operators and sets of axioms. It is proved that hierarchical systems of domain conditions are necessary and sufficient for the definition of substructures by the requirement that the inclusion forms a homomorphism. This implies that for this theory only one concept of homomorphism is necessary to yield a nice theory of structure.

On the basis of the results of the fourth section we introduce in the last section of this chapter hierarchical equationally partial varieties, briefly hep-varieties, and hep-quasi-varieties. The class of small categories turns out to be a typical example of a hep-variety.

2.1 Set-theoretical Notions

As usual we denote sets by capital letters, elements and functions by small letters. For finite sets the elements which belong to the set can also be given within braces.

Let S be any finite set. An S-indexed family $A = (A_s \mid s \in S)$ of sets will briefly be called and S-set. Correspondingly, an S-indexed family $f = (f_s: A_s \longrightarrow B_s \mid s \in S)$ of functions will briefly be called an S-function or S-mapping. If it is evident from context that both A and B are S-sets, we denote an S-function by $f: A \longrightarrow B$. For any $s \in S$ the set A_s and the mapping $f_s: A_s \longrightarrow B_s$ will be called the s-component of the S-set A and of the S-mapping $f: A \longrightarrow B$ respectively.

Let $f: A \longrightarrow B$ be any function and $x \in A$ any element of the domain A of $f: A \longrightarrow B$. The element that is assigned to the element x by the function $f: A \longrightarrow B$ will be denoted by $xf \in B$. If additionally $g: B \longrightarrow C$ is a further mapping, the composition will be denoted by $fg: A \longrightarrow C$, sometimes also by $f \circ g: A \longrightarrow C$. The application of the composition to any $x \in A$ is then denoted by

$$xfg \in C \quad \text{or by} \quad (xf)g \in C$$

if we want to stress the successive application of f and g. This notation is compatible both with the diagrammatic and the applicative order of the composition. We hope that readers will not be confused or irritated by the use of xf instead of $f(x)$.

Very often we have to deal with sets of variables taking values in an S-set $A = (A_s \mid s \in S)$. In this case the particular s-component in which each variable $x \in X$ can take on values must be fixed. Accordingly we define an S-sorted system of variables as a mapping $v: X \longrightarrow S$. Now, any $x \in X$ can take on values in A_{xv} only. An assignment of $v: X \longrightarrow S$ in an S-set $A = (A_s \mid s \in S)$ is therefore given by a function

$$\underline{a}: X \longrightarrow \bigcup_{s \in S} A_s$$

such that $x\underline{a} \in A_{xv}$ for each $x \in X$. A_v denotes the set of all assignments of the S-sorted system of variables $v: X \longrightarrow S$ in the S-set $A = (A_s \mid s \in S)$.

Let $f = (f_s: A_s \longrightarrow B_s \mid s \in S)$ be any S-function and $v: X \longrightarrow S$ any S-sorted system of variables, then we can extend in a natural way the S-function to a function

$$f_v: A_v \longrightarrow B_v$$

by setting $x(\underline{a}f_v) = (x\underline{a})f_{xv} \in B_{xv}$ for every $x \in X$, $\underline{a} \in A_v$.
The mapping $f_v: A_v \longrightarrow B_v$ transforms any assignment $\underline{a} \in A_v$ to an assignment $\underline{a}f_v \in B_v$ of the same S-sorted system of variables $v: X \longrightarrow S$ in the S-set $B = (B_s \mid s \in S)$ by pointwise application of the S-function $f: A \longrightarrow B$.

For the sake of simpler formulae we allow the subscript s and v to be omitted from the functions $f_s: A_s \longrightarrow B_s$, $f_v: A_v \longrightarrow B_v$ respectively, if this subscript can easily be derived from context.

$\mathbb{N} = \{0,1,2,\ldots\}$ denotes the set of natural numbers, and for any $n \in \mathbb{N}$ we set $[n] = \{1,2,\ldots,n\}$ and $[0] = \emptyset$. A finite sorted system of variables $w: [n] \longrightarrow S$ can obviously be identified with the string $w = s_1 s_2 \ldots s_n$ where $s_i = iw$ for every $i \in [n]$. The empty string will be denoted as usual by $\lambda : [0] \longrightarrow S$. S^* denotes the set of all finite strings of elements of S, e.g. the set of all S-sorted systems of variables of the form $w: [n] \longrightarrow S$. S^+ denotes the set of all non-empty strings of elements of S.

For any string $w: [n] \longrightarrow S$ an assignment $\underline{a} \in A_w$ in an S-set $A = (A_s \mid s \in S)$ will also be denoted by

$$\underline{a} = (a_1, a_2, \ldots, a_n) \text{ with } a_i = i\underline{a} \text{ for every } i \in [n]$$

as ordered n-tuples are usually denoted. According to that convention A_w can be identified with the Cartesian product, i.e.

$$A_w = A_{1w} \times A_{2w} \times \ldots \times A_{nw}.$$

We recall that for the empty string $\lambda : [0] \longrightarrow S$ the set A_λ consists of one element for every S-set $A = (A_s \mid s \in S)$ which is the inclusion of the empty set into the union of the s-components of the

S-set. Therefore we can set $A_\lambda = \{\emptyset\}$.

If $w : [n] \longrightarrow S$ is a string, $A = (A_s \mid s \in S)$ is an S-set, $f : A_w \longrightarrow M$ is any function, and if we use the notation $(a_1, \ldots, a_n)$ to denote elements of A_w, then the value associated to $(a_1, \ldots, a_n)$ by $f : A_w \longrightarrow M$ will be denoted by $(a_1, \ldots, a_n)f$.

A binary S-relation ϱ in an S-set $A = (A_s \mid s \in S)$ is given by an S-indexed family $\varrho = (\varrho_s \mid s \in S)$ such that $\varrho_s \subseteq A_s \times A_s$ for every $s \in S$. An S-relation ϱ in an S-set A is refered to as an S-equivalence in A if the binary relation $\varrho_s \subseteq A_s \times A_s$ is an equivalence relation for every $s \in S$, i.e. if ϱ_s is reflexive, transitive and symmetric for every $s \in S$.

Finally we will point out an important effect. Let $v : X \longrightarrow S$ be a sorted system of variables and $A = (A_s \mid s \in S)$ an S-set so that at least one variable $x \in X$ exists with $A_{xv} = \emptyset$. Then there exists no assignment of $v : X \longrightarrow S$ in A such that $A_v = \emptyset$. This effect causes the theory of many-sorted algebras to differ seriously from the theory of homogeneous algebras, especially with respect to the 'equational theory' of many-sorted algebras.

2.2 Many-sorted Algebras

Before we deal with partial algebras we introduce the basic notions of total many-sorted algebras.

Definition 2.2.1 A signature

$$\Sigma = (S, \alpha : \Omega \longrightarrow S^* \times S)$$

is given by a finite set S of so-called sort names, by a finite set Ω of operators or operation symbols, and by an arity function $\alpha : \Omega \longrightarrow S^* \times S$ which associates with every operator $\sigma \in \Omega$ an ordered pair $\sigma\alpha = (w,s)$. The string w is called the domain of σ and $s \in S$ the range of $\sigma \in \Omega$, respectively. Instead of $\sigma = (w,s)$ we often write $\sigma : w \longrightarrow s$.

A Σ-algebra

$$A = ((A_s \mid s \in S), (\sigma^A \mid \sigma \in \Omega))$$

is given by an S-set $(A_s \mid s \in S)$, the underlying S-set of A, and by an Ω-indexed family $(\sigma^A \mid \sigma \in \Omega)$ of fundamental operations, so that for every operator $\sigma \in \Omega$ with $\sigma : w \longrightarrow s$ the associated fundamental operation σ^A is a function from A_w to A_s, i.e.

$$\sigma^A : A_w \longrightarrow A_s .$$

For a Σ-algebra A the underlying S-set of A will also be denoted by A and will be called the S-carrier of the Σ-algebra A.

By $ALG_t(\Sigma)$ we denote the class of all Σ-algebras.

If $A, B \in ALG_t(\Sigma)$, an S-function $f = (f_s : A_s \longrightarrow B_s \mid s \in S)$ is said to be a Σ - h o m o m o r p h i s m from A to B, denoted by $f: A \longrightarrow B$, if for every $\sigma \in \Omega$ with $\sigma : w \longrightarrow s$ and every $\underline{a} \in A_w$ the following equation holds:

$$(\underline{a}\,\sigma^A) f_s = (\underline{a} f_w)\sigma^B ,$$

i.e. the following diagram commutes:

$$\begin{array}{ccc} A_w & \xrightarrow{\quad f_w \quad} & B_w \\ \Big\downarrow \sigma^A & & \Big\downarrow \sigma^B \\ A_s & \xrightarrow{\quad f_s \quad} & B_s \end{array}$$

for every $\sigma \in \Omega$.

A Σ-homomorphism $f: A \longrightarrow B$ is called a Σ-isomorphism, if there is a Σ-homomorphism $g: B \longrightarrow A$ with

$$f_s \cdot g_s = Id_{A_s} , \quad g_s \cdot f_s = Id_{B_s} \quad \text{for every } s \in S . \qquad \blacksquare$$

When we give examples of signatures we will use a more readable representation similar to the specifications of the first chapter.

In the application of algebra in computer science signatures with additional structure serve as mathematical representations of abstract data types, see Chapter I and Chapter IV. Relations between abstract data types and constructions with them will be mathematically described by relations and constructions on structured signatures. For this reason we introduce the notion of a subsignature and of a signature morphism.

Definition 2.2.2. $\Sigma_1 = (S_1, \alpha_1 : \Omega_1 \longrightarrow S_1^* \times S_1)$ is a s u b s i g n a t u r e of $\Sigma_2 = (S_2, \alpha_2 : \Omega_2 \longrightarrow S_2^* \times S_2)$, written $\Sigma_1 \subseteq \Sigma_2$, if $S_1 \subseteq S_2$, $\Omega_1 \subseteq \Omega_2$, and $\sigma\alpha_1 = \sigma\alpha_2$ for every $\sigma \in \Omega_1$.

A s i g n a t u r e m o r p h i s m $(\varphi, \psi) : \Sigma_1 \longrightarrow \Sigma_2$ is

an ordered pair (φ,ψ), where $\varphi: S_1 \longrightarrow S_2$, $\psi: \Omega_1 \longrightarrow \Omega_2$ such that $\psi \cdot \alpha_2 = \alpha_1 \cdot (\varphi^* \times \varphi)$, where $(w,s)(\varphi^* \times \varphi) = (w \cdot \varphi, s)$ for every $w:[n] \longrightarrow S_1$, $s \in S_1$. The string $w \cdot \varphi \in S_2^*$ results from $w:[n] \longrightarrow S_1$ by pointwise application of $\varphi: S_1 \longrightarrow S_2$, i.e. $i(w \cdot \varphi) = (iw)\varphi$ for each $i \in [n]$.
The composition of signature morphism $(\varphi,\psi): \Sigma_1 \longrightarrow \Sigma_2$ and $(\bar{\varphi},\bar{\psi}): \Sigma_2 \longrightarrow \Sigma_3$ is naturally defined by

$$(\varphi,\psi) \circ (\bar{\varphi},\bar{\psi}) = (\varphi \cdot \bar{\varphi}, \psi \cdot \bar{\psi}),$$

i.e. by componentwise composition. ∎

Clearly, $\Sigma_1 \subseteq \Sigma_2$ holds iff (if and only if) the inlusions $S_1 \subseteq S_2$ and $\Omega_1 \subseteq \Omega_2$ form a signature morphism.

We can omit examples of subsignatures and signature morphisms here because they can be found in Chapter I and Chapter IV.

At this point we will draw attention to a construction indispensible for almost all following notions and investigations. Roughly speaking we will make a closer study of all formal expressions which can be constructed from an S-sorted system of variables $v: X \longrightarrow S$ and a signature Σ, and we will consider the external behaviour of this term algebra $T(\Sigma,v)$ with respect to total and partial Σ-algebras. We will see that this external behaviour, described by the following recursion theorem, characterizes $T(\Sigma,v)$ uniquely up to isomorphism. Therefore, not the special construction of $T(\Sigma,v)$ is important but the external behaviour. We will construct $T(\Sigma,v)$ by means of the so-called 'postfix polish notation'.

<u>Definition 2.2.3</u>. Let $\Sigma = (S, \alpha: \Omega \longrightarrow S^* \times S)$ be any signature and $v: X \longrightarrow S$ an S-sorted system of variables.

$$T(\Sigma,v) = (T(\Sigma,v)_s \mid s \in S)$$

is the smallest S-set whose elements are strings from $X \cup \Omega$ (it is supposed that $X \cap \Omega = \emptyset$) satisfying the following conditions:

(1) $x \in T(\Sigma,v)_{xv}$ for every $x \in X$;

(2) $\sigma \in T(\Sigma,v)_s$ for every $\sigma \in \Omega$ and $s \in S$ with $\sigma\alpha = (\lambda, s)$;

(3) If $\sigma \in \Omega$ with $\sigma: w \longrightarrow s$, $w:[n] \longrightarrow S$, $n \neq 0$ and if $\underline{t} = (t_1,\ldots,t_n) \in T(\Sigma,v)_w$ then

$$t_1 t_2 \ldots t_n \sigma \in T(\Sigma,v)_s .$$

In the sequel $t_1t_2...t_n\sigma$ will also be denoted briefly by $\underline{t}\sigma$.
By (2) and (3) we can define easily the fundamental operation

$$\sigma^{T(\Sigma,v)}: T(\Sigma,v)_w \longrightarrow T(\Sigma,v)_s \quad \text{for every} \quad \sigma : w \longrightarrow s \quad \text{in } \Omega.$$

Let $\sigma\alpha = (\lambda,s)$, then

$$\sigma^{T(\Sigma,v)} = \sigma \in T(\Sigma,v)_s$$

and if $\sigma: w \longrightarrow s$, $w: [n] \longrightarrow S$, $n \neq 0$, and $\underline{t} \in T(\Sigma,v)_w$ then

$$\underline{t}\,\sigma^{T(\Sigma,v)} = \underline{t}\sigma \in T(\Sigma,v)_s .$$

By

$$T(\Sigma,v) = ((T(\Sigma,v)_s \mid s \in S),(\sigma^{T(\Sigma,v)} \mid \sigma \in \Omega))$$

we denote the so-called a l g e b r a o f Σ- t e r m s o n $v: X \longrightarrow S$, briefly the Σ-term-algebra on v. ■

This representation of terms will be used only in formal mathematical considerations. When we give examples, we will use a more readable notation with infix notation for binary operators and functional notation with parentheses for other operators in the same way as in Chapter one. If we use terms or term equations to describe properties of operations, then the usual notation is very suitable, but if terms become objects of mathematical investigation another notion is more suitable, at least from our point of view.

Now we can turn to the external behaviour of $T(\Sigma,v)$.

(2.2.4) Recursion Theorem

Let $\Sigma = (S, \alpha : \Omega \longrightarrow S^*\times S)$ be any signature and $v: X \longrightarrow S$ an S-sorted system of variables. For any Σ-algebra A and any assignment $\underline{a} \in A_v$ there exists exactly one homomorphism

$$a: T(\Sigma,v) \longrightarrow A \quad \text{with} \quad xa_{xv} = x\underline{a} \quad \text{for every} \quad x \in X,$$

i.e. every assignment can uniquely be extended to a homomorphism.

Proof:
The uniqueness of the extension to a homomorphism is evident, because $T(\Sigma,v)$ is generated by the S-set $(v^{-1}(s) \mid s \in S)$, where $v^{-1}(s) = \{x \in X \mid xv = s\}$, and because any two homomorphisms which coincide on a generating set are equal. So it remains to prove that $\underline{a} \in A_v$ can be extended to a homomorphism $a: T(\Sigma,v) \longrightarrow A$. By appealing directly to the inductive definition of the S-set of Σ-terms on

$v: X \longrightarrow S$ we can define the homomorphism by $xa_{xv} = x\underline{a}$ for every $x \in X$, $\sigma a_s = \sigma^A$ for every $\sigma \in \Omega$, $s \in S$ with $\sigma : \lambda \longrightarrow s$, and $(\underline{t}\sigma)a_s = (\underline{t}a_w)\sigma^A$ for every $\sigma \in \Omega$, $s \in S$ with $\sigma : w \longrightarrow s$, $w: [n] \longrightarrow S$, $n \neq 0$, and every $\underline{t} \in T(\Sigma, v)_w$.
Obviously, the S-function

$$a = (a_s : T(\Sigma, v)_s \longrightarrow A_s \mid s \in S)$$

is uniquely defined in this way, and it is a homomorphism. ■

We illustrate the recursion theorem by a simple example: Let us consider the signature

Σ __is__ __sorts__ B, C, E
__oprn__ f, t $\longrightarrow$ B
0, 1 $\longrightarrow$ E
$+, \circ$ (E, E) $\longrightarrow$ E
$\wedge, \vee$ (B, B) $\longrightarrow$ B
$\neg$ (B) $\longrightarrow$ B
κ (B, E, E) $\longrightarrow$ C
__end__ Σ

and the sorted system of variables

$$v : \{a, b, c, d, x, y, z\} \longrightarrow \{E, C, B\}$$

with $av = bv = cv = dv = E$, $xv = yv = zv = B$.
Examples of Σ-terms on $v: X \longrightarrow S$ are given by

$tx\wedge\neg y\vee f\wedge \in T(\Sigma, v)_B$,

$ab1{+}{\circ}c{+} \in T(\Sigma, v)_E$,

$dd{\circ}ca{+}{\circ} \in T(\Sigma, v)_E$,

$xy\vee\neg ab{+}b{\circ}abc{\circ}{\circ}1{+}\kappa \in T(\Sigma, v)_C$,

whereas $ax+ \notin T(\Sigma, v)_E$, because $xv = B$ and not $xv = E$, and $abc{+}1{+}$ is also not a Σ-term on $v: X \longrightarrow S$ because one of the binary operators $+, \circ$ is left out after the b or after the c or after the 1 or after the final $+$.

Now let A be the Σ-algebra with $A_E = A_C = \mathbb{N}$, where $\mathbb{N}$ denotes the set of natural numbers, and $A_B = \{T, F\}$, the set of truth values, and where

$t^A = T$, $f^A = F$, $0^A =$ zero, $1^A =$ one,

$+^A =$ addition of natural numbers,

$\circ^A =$ multiplication of natural numbers,

$\wedge^A$ = logical conjunction,

$\vee^A$ = logical disjunction,

$\neg^A$ = logical negation, and

$\kappa^A : \{T,F\} \times \mathbb{N} \times \mathbb{N} \longrightarrow \mathbb{N}$ with

$$(x,n_1,n_2)\kappa^A = \begin{cases} n_1 & \text{if } x = T \\ n_2 & \text{if } x = F . \end{cases}$$

Finally, let be given an assignment $\underline{h} \in A_V$ with $a\underline{h} = 3$, $b\underline{h} = 5$, $c\underline{h} = 2$, $d\underline{h} = 0$, $x\underline{h} = T$, $y\underline{h} = z\underline{h} = F$. The uniquely determined homomorphism

$$h: T(\Sigma,v) \longrightarrow A$$

corresponds, not only in this example, to the evaluation of the formal expressions, here written in postfix polish notation, with respect to the given assignment of the sorted set of variables. According to that

$$(tx\wedge\neg y\vee f\wedge)h_B = ((tx\wedge\neg y\vee)h_B, fh_B)\wedge^A =$$

$$= ((tx\wedge\neg y\vee)h_B, F)\wedge^A = F,$$

$$(ab1+\circ c+)h_E = 3\ (5+1)+2 = 20 ,$$

$$(dd\circ ca+\circ)h_E = ((0\cdot 0)\cdot(2+3)) = 0 ,$$

$$(xy\vee\neg ab+b\circ abc\circ\circ 1+\kappa)h_C = ((xy\vee\neg)h_B, (ab+b\circ)h_E, (abc\circ\circ 1+)h_E)\kappa^A =$$

$$= (\neg(T\vee F), (3+5)\cdot 5, 3\cdot(5\cdot 2)+1)\kappa^A = 3\cdot(5\cdot 2)+1 = 31 .$$

We see that the chosen notation of Σ-terms is not easy to read. If we want to assign to the Σ-terms a more readable notation, i.e. if we want to define another orthography, then we can do it by means of the recursion theorem. For that purpose we define the following 'orthographical Σ-algebra' O, where

$$O_E = \{a,b,c,d,[,],+,0,1\}^+ ,$$

$$O_B = \{x,y,z,\wedge,\vee,\neg,[,],T,F\}^+ ,$$

$$O_C = \{a,b,c,d,x,y,z,+,\wedge,\vee,0,1,T,F,[,],\underline{IF},\underline{THEN},\underline{ELSE},\neg\}^+$$

and where the Σ-indexed family of fundamental operations is given by the following string functions:

$$t^O = T,\ f^O = F,\ 0^O = 0,\ 1^O = 1 ,$$

$$(w_1,w_2)+^O = [w_1+w_2] ,$$

$$(w_1,w_2)\circ^O = w_1w_2 ,$$

$$(w_1,w_2)\vee^O = [w_1\vee w_2] ,$$

$(w_1,w_2)\wedge^O = [w_1 \wedge w_2]$,

$w \neg^O = \neg[w]$,

$(w_1,w_2,w_3)\kappa^O = \underline{IF}\ w_1\ \underline{THEN}\ w_2\ \underline{ELSE}\ w_3$.

The last string function κ^O acts for instance as follows: write the character $\underline{IF}$, append the first argument, append the character $\underline{THEN}$ append the second argument, append the character $\underline{ELSE}$ and append finally the third argument. If we choose as assignment $\underline{i} \in O_V$ the inclusion, then the homomorphism

$$i: T(\Sigma,v) \longrightarrow O$$

associates with the terms their new notation, so for instance

$$(xy\vee\neg ab{+}b{\cdot}abc\circ\circ 1{+}\kappa)i_C = \underline{IF}\neg[x\vee y]\underline{THEN}[a{+}b]b\ ELSE[abc{+}1] .$$

But, not every string of the carrier of O is an image with respect to the homomorphism $i: T(\Sigma,v) \longrightarrow O$. In addition, one can see that different formal expressions can be mapped to the same notation, so for instance

$$(abc\circ\circ)i_E = (ab\circ c\circ)i_E .$$

By the preceding example we can illustrate one of the main problems of programming languages, namely the definition of syntax and semantics. With respect to our example we can say that the chosen orthography, represented by the Σ-algebra O, and the defined semantics, represented by the Σ-algebra A, are consistent, if $t_1 i_s = t_2 i_s$ for $s \in \{B,E,C\}$, $t_1,t_2 \in T(\Sigma,v)_s$ implies $t_1 h_s = t_2 h_s$ for every homomorphism

$$h: T(\Sigma,v) \longrightarrow A$$

extending an assignment $\underline{h} \in A_V$, i.e. equal notation of formal expressions implies equal meaning. Precisely this principle was used in Section 1.3 for the definition of the semantics of the functional language.

Up to now we have only defined the meaning of a formal expression $t \in T(\Sigma,v)$ and not yet the meaning of an orthographical correct notation, i.e. of an element $w \in O_s$, $s \in \{B,C,E\}$, that is an image of a formal expression with respect to the homomorphism $i: T(\Sigma,v) \longrightarrow O$. The answer to the last question can be given by the well-known theorem of homomorphisms. To formulate this theorem we need some further basic algebraic notions.

Definition 2.2.5. Let $\Sigma = (S, \alpha : \Omega \longrightarrow S^* \times S)$ be any signature. A Σ-algebra A is called a s u b a l g e b r a of a Σ-algebra B if

(1) $A_s \subseteq B_s$ for every $s \in S$, and

(2) the S-inclusion $i: A \longrightarrow B$, i.e. $xi_s = x$ for every $s \in S$, $x \in A_s$, is a Σ-homomorphism. ∎

Obviously, for any S-subset $A \subseteq B$ of the underlying S-set of a Σ-algebra B there is at most one Σ-algebra with A as underlying S-set and which is a Σ-subalgebra of B. An S-subset $A \subseteq B$ of the S-carrier of a Σ-algebra B is an S-carrier of a Σ-subalgebra if and only if for every $\sigma \in \Omega$, where $\sigma : w \longrightarrow s$, and for every $\underline{a} \in A_w$ the application of σ^B to $\underline{a}$ yields an element of A_s, i.e. A is closed with respect to the application of fundamental operations of B to arguments taken from A.

For the sake of completeness we summarize here some well-known facts of the theory of total many-sorted algebras without proof. Proofs and more details can be found in the book of Lugowski (1976) or in the paper of Birkhoff and Lipson (1970).

(2.2.6) Let $A, B \in ALG_t(\Sigma)$. For any Σ-homomorphism $f: A \longrightarrow B$ the set-image $Af \subseteq B$ of the S-set A is an S-carrier of a Σ-subalgebra of B, which will be called the h o m o m o r p h i c i m a g e, denoted by Af. ∎

According to (2.2.6) the $\{B,E,C\}$-set of orthographically correct representations of formal expressions is a carrier of a subalgebra of O.

(2.2.7) For any $A \in ALG_t(\Sigma)$ an S-indexed family $\varrho = (\varrho_s \mid s \in S)$ of equivalence relations $\varrho_s \subseteq A_s \times A_s$, $s \in S$, is called a c o n g r u e n c e r e l a t i o n in A if $(\underline{x}\sigma^A, \underline{y}\sigma^A) \in \varrho_s$ holds for every $\sigma \in \Omega$, $\sigma : w \longrightarrow s$, and for every $\underline{x}, \underline{y} \in A_w$ with $(i\underline{x}, i\underline{y}) \in \varrho_{iw}$ for every $i \in [n]$, where $w: [n] \longrightarrow S$. ∎

(2.2.8) For any Σ-homomorphism $f: A \longrightarrow B$ the S-indexed family

$$\ker f = (\ker f_s \mid s \in S) \text{ with}$$

$$\ker f_s = \{(x,y) \in A_s \times A_s \mid xf_s = yf_s\}, \ s \in S,$$

is a congruence relation in A. ∎

(2.2.9) Theorem of Homomorphisms

Let be given Σ-homomorphisms $f: B \longrightarrow A$, $g: B \longrightarrow C$, so that $g_s: B_s \longrightarrow C_s$ is surjective for every $s \in S$, briefly $g: B \longrightarrow C$ is surjective. Then there exists a Σ-homomorphism $h: C \longrightarrow A$ with $f = g \circ h$ if and only if $\ker g_s \subseteq \ker f_s$ for every sort name $s \in S$. ∎

The homomorphism theorem can be applied to the problem described above if we take $B = T(\Sigma,v)$, for A the semantic algebra, for C the homomorphic image $T(\Sigma,v)i$ of $i: T(\Sigma,v) \longrightarrow O$, i.e. the subalgebra of orthographically correct representations of formal expressions, and if we assume that $f: T(\Sigma,v) \longrightarrow A$ is the extension of any assignment $\underline{f} \in A_v$, and that $g: T(\Sigma,v) \longrightarrow T(\Sigma,v)i$ is the restriction of $i: T(\Sigma,v) \longrightarrow O$ to its homomorphic image. The condition that equal representations imply equal meaning guarantees then by (2.2.9) the existence of a homomorphism $h: T(\Sigma,v)i \longrightarrow A$ which defines the meaning for every orthographically correct representation.

The logical structure of a compiler can be reconstructed from this algebraic definition of semantics by $h: T(\Sigma,v)i \longrightarrow A$. The aim of syntacic analysis is to decide whether a given string is an orthographically correct representation of a formal expression and if so to construct the corresponding internal representation. Sometimes even the postfix polish notation is used as the internal form of formal expressions. The second part of a compiler, the semantic interpretation, corresponds to a description of the semantic algebra in terms of the machine or assembler language of the target computer. The main point of this description is the definition of the fundamental operations of A by suitable programs.

Now that this example is complete we shall make a closer study of the relations between $T(\Sigma,v)$ and partial Σ-algebras for a general signature Σ and any S-sorted system of variables $v: X \longrightarrow S$.

Definition 2.2.10. For any signature $\Sigma = (S, \alpha: \Omega \longrightarrow S^* \times S)$ a p a r t i a l Σ-a l g e b r a

$$A = ((A_s \mid s \in S),(\sigma^A \mid \sigma \in \Omega))$$

is given by an S-set A and by a Σ-indexed family of partial fundamental operations, so that the domain of the partial fundamental operation σ^A, denoted by $\operatorname{dom} \sigma^A$, is a subset of A_w if $\sigma: w \longrightarrow s$, i.e.

$$\sigma^A: \operatorname{dom} \sigma^A \longrightarrow A_s \quad \text{and} \quad \operatorname{dom} \sigma^A \subseteq A_w .$$

By ALG(Σ) we denote the class of all partial Σ-algebras.
If $A, B \in$ ALG(Σ), an S-mapping $f = (f_s : A_s \longrightarrow B_s \mid s \in S)$ is called a Σ-homomorphism from A to B, denoted by $f: A \longrightarrow B$, if for every $\sigma \in \Omega$, with $\sigma : w \longrightarrow s$ and every $\underline{a} \in A_w$ the following conditions hold:

(1) $\underline{a} \in \mathrm{dom}\, \sigma^A$ implies $\underline{a} f_w \in \mathrm{dom}\, \sigma^B$.

(2) $\underline{a} \sigma^A f_s = \underline{a} f_w \sigma^B$ if both sides are defined.

A Σ-homomorphism $f: A \longrightarrow B$ between partial Σ-algebras is called c l o s e d (strong in the sense of Grätzer (1978)) if additionally

(3) $\underline{a} f_w \in \mathrm{dom}\, \sigma^B$ implies $\underline{a} \in \mathrm{dom}\, \sigma^A$ for every $\underline{a} \in A_w$. ∎

The various kinds of homomorphisms between partial algebras are extensively described by Burmeister in Part I. We have introduced here only the weakest and the strongest versions of homomorphism, because they are the only ones we need for the concept of equational partiality. Since we are not interested in arbitrary partial algebras, we give no examples for (2.2.10) here. Readers who are interested in this direction should consult the book of Burmeister.

In the second half of this section we introduce the concept of 'existence equations for many-sorted algebras'. For the case of homogeneous algebras this concept is discussed by Burmeister in detail, so that we give here only a brief but complete introduction.

(2.2.11) Let $v: X \longrightarrow S$ be any S-sorted system of variables and $\Sigma = (S, \alpha : \Omega \longrightarrow S^* \times S)$ any signature. If $\sigma \in \Omega$, $\sigma : w \longrightarrow s$, $w: [n] \longrightarrow S$, and

$$t = \underline{t}\, \sigma^{T(\Sigma, v)} \quad \text{for some } \underline{t} \in T(\Sigma, v)_w$$

then every term $i\underline{t}$ for $i \in [n]$ is called an i m m e d i a t e s u b t e r m of $t \in T(\Sigma, v)_s$. We denote by $\leq$ the reflexive and transitive closure of the immediate subterm relation, and call it the subterm relation. ∎

It is obvious that the subterm relation is a partial order on the union of the carriers of $T(\Sigma, v)$. The minimal elements of this poset are the variables and the constant operators. Additionally, the poset of all Σ-terms on $v: X \longrightarrow S$ satisfies the decreasing chain condition, that means, any strictly decreasing chain

$$t_0 > t_1 > \ldots > t_n$$

has finite length.

(2.2.12) Let be given $\Sigma = (S, \alpha : \Omega \longrightarrow S^* \times S)$, $v: X \longrightarrow S$, a partial Σ-algebra $A \in ALG(\Sigma)$, and an assignment $\underline{a} \in A_v$.
Then we define inductively the S-set

$$\mathrm{dom}\,\underline{\tilde{a}} \subseteq T(\Sigma, v)$$

of all $\underline{a}$ - i n t e r p r e t a b l e terms, and define also inductively the value

$$t\underline{\tilde{a}}_s \in A_s \quad \text{of the } \underline{a}\text{-interpretation of} \quad t \in T(\Sigma, v)_s$$

as follows:

(1) Every variable $x \in X$ is $\underline{a}$-interpretable, and $x\underline{\tilde{a}}_{xv} = x\underline{a}$ for every $x \in X$.

(2) Let $\sigma \in \Omega$, $\sigma : w \longrightarrow s$, $w: [n] \longrightarrow S$, $\underline{t} \in T(\Sigma, v)_w$ and $t = \underline{t}\sigma^{T(\Sigma, v)}$.
Then t is $\underline{a}$-interpretable if and only if firstly $i\underline{t}$ is $\underline{a}$-interpretable for every $i \in [n]$ and secondly $\underline{t}\underline{\tilde{a}}_w \in \mathrm{dom}\,\sigma^A$.
The $\underline{a}$-interpretation of t is then defined by

$$(\underline{t}\sigma^{T(\Sigma, v)})\underline{\tilde{a}}_s = (\underline{t}\underline{\tilde{a}}_w)\sigma^A . \qquad \blacksquare$$

If $A \in ALG_t(\Sigma)$, then any term is $\underline{a}$-interpretable and the unique extension of $\underline{a} \in A_v$ to a homomorphism coincides with the $\underline{a}$-interpretation

$$\underline{\tilde{a}} : \mathrm{dom}\,\underline{\tilde{a}} = T(\Sigma, v) \longrightarrow A .$$

It is easy to see that the set of $\underline{a}$-interpretable terms is closed with respect to subterms. Furthermore, we mention that the homomorphism

$$\underline{\tilde{a}} : \mathrm{dom}\,\underline{\tilde{a}} \longrightarrow A$$

is a closed homomorphism, and that the $\underline{a}$-interpretation is the greatest extension of $\underline{a} \in A_v$ to a closed homomorphism

$$\underline{a}^{\#} : B \longrightarrow A ,$$

where B is a partial Σ-algebra such that $B_s \subseteq T(\Sigma, v)_s$ for every $s \in S$, where the S-inclusion forms a homomorphism and the union of the s-carriers of B is closed with respect to subterms.

Definition 2.2.13. An ordered pair $(t_1, t_2) \in T(\Sigma, v)_s \times T(\Sigma, v)_s$, $s \in S$, of Σ-terms on $v: X \longrightarrow S$ is said to be an e x i s t e n c e e q u a t i o n on $v: X \longrightarrow S$, denoted by

$$(v: t_1 \stackrel{e}{=} t_2) .$$

If A is any partial Σ-algebra, an assignment $\underline{a} \in A_v$ is called a s o l u t i o n of $(v: t_1 \stackrel{e}{=} t_2)$ in A, if

$$t_1 \in (\mathrm{dom}\,\underline{\tilde{a}})_s, \quad t_2 \in (\mathrm{dom}\,\underline{\tilde{a}})_s \quad \text{and} \quad t_1\underline{\tilde{a}}_s = t_2\underline{\tilde{a}}_s .$$

This fact will be denoted by

$$(A,\underline{a}) \models (v: t_1 \stackrel{e}{=} t_2) .$$

By $(v:G)$ where

$$G \subseteq \bigcup_{s \in S} T(\Sigma,v)_s \times T(\Sigma,v)_s ,$$

we denote a set of existence equations on $v: X \longrightarrow S$.

For a set of existence equations $(v:G)$ and any partial Σ-algebra A we define the s e t o f s o l u t i o n s of $(v:G)$ in A by

$$A_{(v:G)} = \{\underline{a} \in A_v \mid (A,\underline{a}) \models (v: t_1 \stackrel{e}{=} t_2) \text{ for all } (v: t_1 \stackrel{e}{=} t_2) \in G\},$$

$$A_{(v:\emptyset)} = A_v .$$

■

It is useful to mention here that the concept of existence equations is not powerful enough to describe the interesting laws interrelating partial operations. This is not caused by a wrong concept of validity of equations for partial algebras. Up to now no concept of validity of equations is known which is capable of defining the class of all small categories as a class of equationally definable partial algebras.

We illustrate the concept of $\underline{a}$-interpretation in partial algebras on the basis of the following partial algebra M for the example of a signature defined above:

$M_B = \{T,F\}$ the set of truth values;

$M_E = M_C$ = set of all (n,k)-matrices, where $n,k \in \mathbb{N}$ are not fixed, with natural numbers as elements;

$t^M = T$, $f^M = F$, $0^M = (0)_{1,1}$, $1^M = \begin{pmatrix} 1 & 0 \\ 0 & 1 \end{pmatrix}_{2,2}$,

$\wedge^M$ = logical conjunction;

$\vee^M$ = logical disjunction;

$\neg^M$ = logical negation;

$+^M$ = addition of matrices;

$\circ^M$ = multiplication of matrices;

$$(x,m_1,m_2)\kappa^M = \begin{cases} m_1 & \text{if } x = T \\ m_2 & \text{if } x = F \end{cases} \quad \text{for all } m_1,m_2 \in M_E = M_C .$$

An assignment $\underline{h} \in M_V$ may be given by $x\underline{h} = y\underline{h} = T$, $z\underline{h} = F$, $a\underline{h}$ = any $(1,2)$-matrix, $b\underline{h}$ = any $(3,4)$-matrix, $c\underline{h}$ = any $(5,5)$-matrix, and $d\underline{h}$ = any $(2,3)$-matrix.

Now the binary operations $+^M$, $\circ^M$ are partial ones, because an (n_1,k_1)-matrix m_1 can be added to an (n_2,k_2)-matrix m_2 iff $n_1 = n_2$ and $k_1 = k_2$, and the product $(m_1,m_2)\circ^M$ exists iff $n_2 = k_1$.

It is easy to see that the following formal expressions are $\underline{h}$-interpretable in M : 11°, 11·1∘d°, db°, db∘db∘+, ccc++c°, whereas the following formal expressions are not $\underline{h}$-interpretable in M : 01+, bc+, bc°, bd°.

Another notion, important for technical reasons, is the s i m u l t a n e o u s s u b s t i t u t i o n o f v a r i a b l e s in a formal expression by given formal expressions. To be more precise, let be given S-sorted systems of variables $x: X \longrightarrow S$, $y: Y \longrightarrow S$, $z: Z \longrightarrow S$ and term assignments $\underline{t}_1 \in T(\Sigma, y)_x$, $\underline{t}_2 \in T(\Sigma, z)_y$. We make the situation clearer by the following diagram:

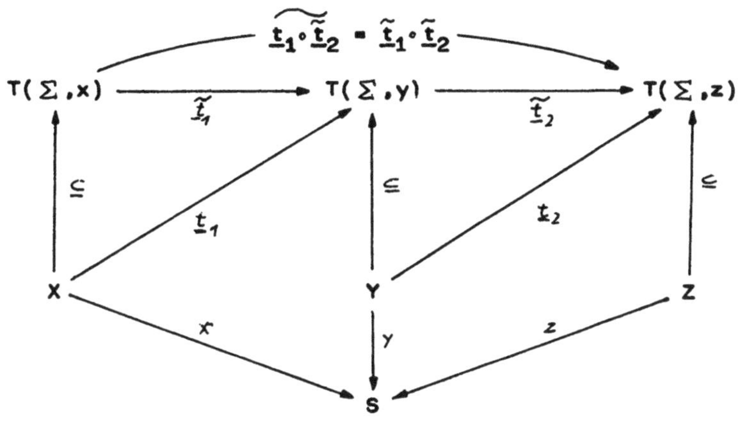

Since $T(\Sigma, y)$, $T(\Sigma, z)$ are total Σ-algebras $\operatorname{dom} \underline{t}_1 = T(\Sigma, x)$ and $\operatorname{dom} \underline{t}_2 = T(\Sigma, y)$. The assignment

$$\underline{t}_1 \circ \tilde{\underline{t}}_2 \in T(\Sigma, z)_x$$

results from $\underline{t}_1$ by simultaneous substitution of the variables contained in $x\underline{t}_1 \in T(\Sigma, y)$, $x \in X$, according to the assignment $\underline{t}_2 \in T(\Sigma, z)_y$. Additionally holds

$$\widetilde{\underline{t}_1 \circ \tilde{\underline{t}}_2} = \tilde{\underline{t}}_1 \circ \tilde{\underline{t}}_2 .$$

We extend this notion from terms and term assignments to existence equations and sets of existence equations. Given an existence equation $(v: t_1 \overset{e}{=} t_2)$ or a set of existence equations $(v: G)$, and an assignment $\underline{t} \in T(\Sigma, w)_v$ then

$$(v: t_1 \overset{e}{=} t_2)\tilde{\underline{t}} = (w: t_1\tilde{\underline{t}} \overset{e}{=} t_2\tilde{\underline{t}}),$$

$$(v: \emptyset)\tilde{\underline{t}} = (w: \emptyset),$$

$$(v:G)\underline{\tilde{t}} = \{ (v: t_1 \overset{e}{=} t_2)\underline{\tilde{t}} \mid (v: t_1 \overset{e}{=} t_2) \in G \} =$$

$$= \{ (w: t_1\underline{\tilde{t}} \overset{e}{=} t_2\underline{\tilde{t}}) \mid (v: t_1 \overset{e}{=} t_2) \in G \}.$$

We conclude this section by a further technical notion. For any term $t \in T(\Sigma,v)$, $v: X \longrightarrow S$, and any set of existence equations $(v:G)$ we define the sets of operators

$$op(t) \subseteq \Omega \quad \text{and} \quad op(v:G) \subseteq \Omega$$

as the sets of operators syntactically contained in t and in $(v:G)$ respectively:

$op(x) = \emptyset$ for every $x \in X$,

$op(\underline{t}) = \bigcup_{i \in [n]} op(i\underline{t})$ for $\underline{t} \in T(\Sigma,v)_w$, $w: [n] \longrightarrow S$,

$op(t) = op(\underline{t}) \cup \{6\}$, for $t = \underline{t}\,6^{T(\Sigma,v)}$ with $6 \in \Omega$, $6: w \longrightarrow s$, $\underline{t} \in T(\Sigma,v)_w$,

$op(\underline{t}) = \bigcup_{y \in Y} op(y\underline{t})$ for $\underline{t} \in T(\Sigma,v)_w$, $w: Y \longrightarrow S$,

$op(v: t_1 \overset{e}{=} t_2) = op(t_1) \cup op(t_2)$,

$op(v:G) = \cup \{ op(v: t_1 \overset{e}{=} t_2) \mid (v: t_1 \overset{e}{=} t_2) \in G \}$.

2.3 Equationally partial heterogeneous algebras

As mentioned above, we describe the partiality by so-called domain conditions, associated with every operator which are necessary and sufficient for the applicability of the operator. Recall the example M of Section 2.2 where $+^M$ is the addition of matrices and $\circ^M$ is the multiplication of matrices with natural numbers as elements. It is well known that both $+^M$ and $\circ^M$ have a necessary and sufficient domain condition:

$(m_1,m_2)+^M$ is defined iff the number of rows and columns of the matrices m_1 and m_2 are equal, and

$(m_1,m_2)\circ^M$ is defined iff the number of rows of m_1 equals the number of columns of m_2.

But these domain conditions are formulated by means of natural numbers. Does this mean that the concept of natural numbers has to be included into the theory of partial algebras? One should be very careful when enriching the theory of partial algebras with any additional tools. The theory should be technically as simple as possible. One can expect that other examples would require other additional concepts if one wants to formulate the usual domain conditions directly.

Therefore we look for another formulation of the domain condition of $+^M$ and $\circ^M$. The concept of natural numbers can be reformulated in terms of matrices themselves using the unit matrices $(1)_{n,n}$. So, we can define additional unary operations

$$1^M : M_E \longrightarrow M_E, \quad r^M : M_E \longrightarrow M_E$$

such that for any (n,k)-matrix $m \in M_E$

$$m\,1^M = \text{(n,n)-unit-matrix},$$

$$m\,r^M = \text{(k,k)-unit-matrix}.$$

Using these operations we can reformulate the domain conditions as follows

$$(m_1,m_2)\,+^M \text{ is defined iff } m_1 1^M = m_2 1^M, \text{ and } m_1 r^M = m_2 r^M,$$

$$(m_1,m_2)\,\circ^M \text{ is defined iff } m_1 r^M = m_2 1^M.$$

We draw attention to the fact that now term equations serve as domain conditions.

Readers familiar with category theory will not be surprised by this fact, since the algebra M is derived from some category of linear mappings.

For small categories the domain condition of the partial binary multiplication (i.e. composition of morphisms) is a term equation, too.

Appealing to these examples, we adopt the convention that domain conditions are always sets of existence equations.

Definition 2.3.1. An e q u a t i o n a l l y p a r t i a l s i g n a t u r e, briefly ep-signature

$$\Theta = (S, \alpha : \Omega \longrightarrow S^* \times S, (\mathrm{def}\,\sigma \mid \sigma \in \Omega))$$

is given by an operator scheme, i.e. by a signature $(S, \alpha : \Omega \longrightarrow S^* \times S)$, and a family $(\mathrm{def}\,\sigma \mid \sigma \in \Omega)$ of sets of existence equations, so that for every $\sigma \in \Omega$ where $\sigma : w \longrightarrow s$, the domain condition $\mathrm{def}\,\sigma$ is a set of existence equations on $w \in S^*$. A partial $(S, \alpha : \Omega \longrightarrow S^* \times S)$-algebra

$$A = ((A_s \mid s \in S), (\sigma^A \mid \sigma \in \Omega))$$

is said to be an equationally partial many-sorted algebra of type Θ, briefly Θ - a l g e b r a , if

$$\operatorname{dom} \sigma^A = A_{\operatorname{def} \sigma}$$

for every $\sigma \in \Omega$.

It is very useful to extend the notion of domain conditions from operators to terms. For that reason we extend the notion $\sigma : w \longrightarrow s$ from operators to terms, too. If $v: X \longrightarrow S$ and $w: Y \longrightarrow S$ are S-sorted systems of variables, then

$t: w \longrightarrow s$ abbreviates $t \in T(\Sigma, w)_s$, and

$\underline{t}: w \longrightarrow v$ abbreviates $\underline{t} \in T(\Sigma, w)_v$.

(2.3.2) Domain conditions of terms can be defined by induction as follows:

(a) $\operatorname{def} x = (v:\emptyset)$ for every $x \in X$, where $v: X \longrightarrow S$ and $x : v \longrightarrow xv$ is considered as a term;

(b) If $\sigma \in \Omega$, $\sigma : \lambda \longrightarrow s$ then

$$\operatorname{def} \sigma^{T(\Sigma, v)} = (\operatorname{def} \sigma)\tilde{\underline{t}}$$

where $\underline{t} : v \longrightarrow \lambda$ is the only assignment of the empty string in $T(\Sigma, v)$;

(c) If $\sigma \in \Omega$, $\sigma : w \longrightarrow s$, $w: [n] \longrightarrow S$, $n \neq 0$, $\underline{t}: v \longrightarrow w$ and $t = \underline{t}\sigma^{T(\Sigma, v)}$, then

$$\operatorname{def} t = (\operatorname{def} \sigma)\tilde{\underline{t}} \cup \bigcup_{i \in [n]} \operatorname{def} i\underline{t} .$$

For any existence equation $(v: t_1 \overset{e}{=} t_2)$ we define its domain condition $\operatorname{def}(v: t_1 \overset{e}{=} t_2)$ as the union of the domain conditions of t_1 and t_2. For any $(v:G)$ we define its domain condition $\operatorname{def}(v:G)$ as the union of the domain conditions of its elements. Finally we set $\operatorname{def}(v:\emptyset) = (v:\emptyset)$.

It is not only useful to extend notions concerning operators to terms but most properties of operations can be generalized inductively to terms. We underline this remark by the following corollary. To formulate this corollary we have to extend the interpretation of operators by their defining operations to terms by so-called t e r m f u n c t i o n s . Thus, we associate with every term $t: v \longrightarrow s$, where $v: X \longrightarrow S$ and $s \in S$, and every partial Σ-algebra A a term function

$$t^A: \operatorname{dom} t^A \longrightarrow A_s \quad \text{with} \quad \operatorname{dom} t^A \subseteq A_v .$$

(2.3.3): Let be given $\Theta = (\Sigma, (\mathrm{def}\,\sigma \mid \sigma \in \Omega))$, $v\colon X \longrightarrow S$ and $A \in ALG(\Sigma)$. Then we define:

(a) With any $x \in T(\Sigma,v)_{xv}$ we associate

$$x^A : A_v \longrightarrow A_{xv}$$

so that $\underline{a}(x^A) = x\underline{a}$ for every $\underline{a} \in A_v$.

(b) $\sigma : \lambda \longrightarrow s$. Although in this case the fundamental operation σ^A and the associated terms function are different mappings, we will use the same notation. It is always clear from context, whether σ^A denotes the fundamental operation of A or the term function. The mappings are different only in their domains. Now we define the terms function $\sigma^A\colon \mathrm{dom}\,\sigma^A \longrightarrow A_s$. We set $\mathrm{dom}\,\sigma^A = \emptyset$ if the fundamental operation σ^A does not exist, and if σ^A is defined in A we set $\mathrm{dom}\,\sigma^A = A_v = \{\emptyset\}$ and $\underline{a}\,\sigma^A = \sigma^A$.

(c) Let be $\sigma \in \Omega$, $\sigma : w \longrightarrow s$, $w\colon [n] \longrightarrow S$, $n \neq 0$, $\underline{t} \in T(\Sigma,v)_w$ and $t = \underline{t}\,\sigma^{T(\Sigma,v)}$. Then we set

$$\mathrm{dom}\,t^A = \{\underline{a} \in A_v \mid \underline{a} \in \mathrm{dom}(i\underline{t})^A \text{ for every } i \in [n] \text{ and } (\underline{a}(1\underline{t})^A,\ldots,\underline{a}(n\underline{t})^A) \in \mathrm{dom}\,\sigma^A\}$$ and

$$\underline{a}\,t^A = (\underline{a}(1\underline{t})^A,\ldots,\underline{a}(n\underline{t})^A)\,\sigma^A$$ ∎

By a very simple induction one can prove

(2.3.4)

(a) $\mathrm{dom}\,t^A = \{\underline{a} \in A_v \mid t \in \mathrm{dom}\,\tilde{\underline{a}}\}$

for every $t \in T(\Sigma,v)_s$, $v\colon X \longrightarrow S$, and $s \in S$.

(b) $\mathrm{dom}\,\tilde{\underline{a}} \subseteq \mathrm{dom}\,\widetilde{\underline{a}f_v}$ and

$(t\tilde{\underline{a}}_s)\,f_s = t(\widetilde{\underline{a}f_v})_s$

for every $t \in T(\Sigma,v)_s$, $v\colon X \longrightarrow S$, $s \in S$, $\underline{a} \in \mathrm{dom}\,t^A$ and every homomorphism $f\colon A \longrightarrow B$. ∎

The conditions of (2.3.4.b) can be visualized by the following diagram:

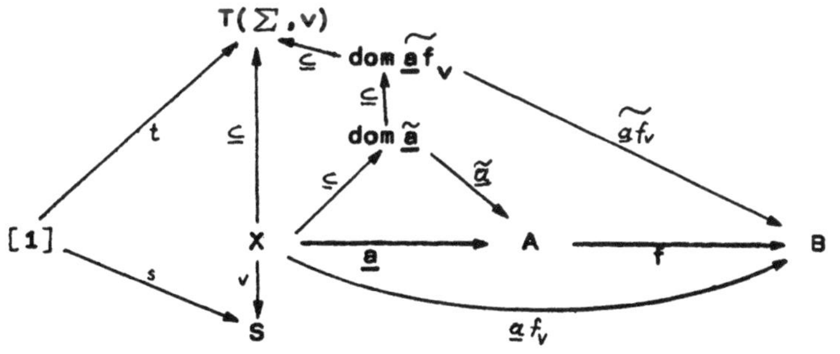

The meaning of a term function t^A can more easily be derived from (2.3.3), but for dealing with term functions in proofs the representation in (2.3.4) is of great value.

By the following corollary we state some properties of term functions.

Corollary 2.3.5.

(a) A partial Σ-algebra $A \in ALG(\Sigma)$ is an equationally partial algebra $A \in ALG(\Sigma, (def\, \sigma \mid \sigma \in \Omega))$, if and only if

$$dom\, t^A = A_{def\, t}$$

for every term $t \in T(\Sigma, v)$, $v: X \longrightarrow S$.

(b) If $A, B \in ALG(\Sigma)$, $f = (f_s: A_s \longrightarrow B_s \mid s \in S)$, then f is a homomorphism $f: A \longrightarrow B$, if and only if for every $v: X \longrightarrow S$, $t \in T(\Sigma, v)$, $\underline{a} \in A_v$ it holds that $\underline{a} \in dom\, t^A$ implies $\underline{a}\, f_v \in dom\, t^B$ and

$$(\underline{a}\, t^A) f_s = (\underline{a}\, f_v)\, t^B .$$

Proof.

At first we assume $A \in ALG(\Sigma, (def\, \sigma \mid \sigma \in \Omega))$ and show $dom\, t^A = A_{def\, t}$ inductively. Therefore, let $t = x$. Then $dom\, t^A = A_v$ holds according to (2.3.3.a), and because $def\, x = (v:\emptyset)$ we obtain $dom\, t^A = A_v = A_{(v:\emptyset)} = A_{def\, t}$. Now let $t = \underline{t}\, \sigma^{T(\Sigma,v)}$ and $dom(i\underline{t})^A = A_{def\, i\underline{t}}$ for every $i \in [n]$. If $\underline{a} \in dom\, t^A$ then we can infer as follows according to (2.3.3.b and c) and by means of Section 2.2:

$$dom\, t^A = \bigcap_{i \in [n]} dom(i\underline{t})^A \cap \{ \underline{a} \in A_v \mid \underline{t}\, \underline{a}_w \in dom\, \sigma^A \} =$$

$$= \bigcap_{i \in [n]} A_{def\, i\underline{t}} \cap \{ \underline{a} \in A_v \mid \underline{t}\, \underline{a}_w \in A_{def\, \sigma} \} =$$

$$= \bigcap_{i \in [n]} A_{def\, i\underline{t}} \cap \{ \underline{a} \in A_v \mid \underline{a} \in A_{(def\, \sigma)\underline{t}} \} =$$

$$= A_{(\bigcup_{i \in [n]} def\, i\underline{t}\ \cup\ (def\, \sigma)\, \underline{t}\,)} = A_{def\, t} .$$

The converse direction is evident since every operator $\sigma \in \Omega$ with $\sigma: w \longrightarrow s$, $w: [n] \longrightarrow S$ can be identified with the term

$$\underline{x}\, \sigma^{T(\Sigma, v)} ,$$

where $i\underline{x} = i$ for every $i \in [n]$. With the polish postfix notation of terms we get

$$\underline{x}\, \sigma^{T(\Sigma, v)} = 12\ldots n\, \sigma \quad \text{and} \quad (12\ldots n\, \sigma)^A = \sigma^A .$$

Secondly we assume that $f: A \longrightarrow B$ is a homomorphism. Because of (2.3.4) $\underline{a} \in \mathrm{dom}\, t^A$ yields $t \in \mathrm{dom}\, \tilde{\underline{a}} \subseteq \mathrm{dom}\, \widetilde{\underline{a}f_V}$, so that $\underline{a}f_V \in \mathrm{dom}\, t^B$, and $(\underline{a}\, t^A)\, f_s = (t\, \tilde{\underline{a}}_s)\, f_s = t(\tilde{\underline{a}}_s\; f_s) = t(\widetilde{\underline{a}\, f_V})_s = (\underline{a}\, f_V)t^B$. ∎

Now we are in a position to continue the concept of equational partiality.

For this reason we consider the following ep-signature

$$\Theta_o = (\Sigma, (\mathrm{def}\, \sigma \mid \sigma \in \Omega))$$

such that $\mathrm{def}\, \sigma = (w: \{12\ldots n\sigma \overset{e}{=} 12\ldots n\sigma\})$ for every $\sigma \in \Omega$, where $\sigma : w \longrightarrow s$, $w: [n] \longrightarrow S$.

Clearly, every partial Σ-algebra is a Θ_o-algebra, too.

If $A \in \mathrm{ALG}(\Sigma)$, $\sigma \in \Omega$ with $\sigma : w \longrightarrow s$ then

$$^A(w: \{12\ldots n\sigma \overset{e}{=} 12\ldots n\sigma\}) = \{\underline{a} \in A_w \mid \underline{a} \in \mathrm{dom}\, \sigma^A\}$$

so this kind of domain condition implies no restriction whatsoever.

This example shows that the concept of equational partiality as developed up to this point does not exclude the problems and difficulties of unrestricted partial algebras investigated in Part I of Introduction to Theory and Application of Partial Algebras by P. Burmeister.

By the next example we show that the concept of a Θ-subalgebra cannot be reduced to the requirement that the inclusion is a homomorphism. We will show that on one and the same S-subset non-isomorphic Θ-subalgebras exist.

In connection with this example we demonstrate our convention for representing equationally partial operators.

Θ_1 <u>is</u> <u>sorts</u> M
<u>oprn</u> k $\longrightarrow$ M
f(x:M <u>iff</u> xg=k) $\longrightarrow$ M
g(x:M <u>iff</u> xf=k) $\longrightarrow$ M
<u>end</u> Θ_1

We consider the Θ_1-algebras A, B with

$A_M = B_M = \{a\}$,

$k^A = k^B = a$, $f^B = g^B = \mathrm{Id}_{\{a\}}$, and $\mathrm{dom}\, f^A = \mathrm{dom}\, g^A = \emptyset$.

It is evident that the total algebra B is really a Θ_1-algebra.

The set of solutions in A of the domain condition

$$\text{def}\, f = (w:\{x\} \longrightarrow \{M\} : \{xg \overset{e}{=} k\})$$

consists of all those elements $a \in A_M$ such that ag^A exists and equals $a = k^A \in A_M$. Due to $\text{dom}\, g^A = \emptyset$ the set of solutions of def f in A is empty, too. In the same way follows $A_{\text{def}\, g} = \emptyset$. Therefore A is also a Θ_1-algebra. Both A and B are subalgebras of the Θ_1-algebra C given by

$$C_M = \{a,b\}\ ,\ k^C = a\ ,\ f^C = g^C : \{a,b\} \longrightarrow \{a,b\} \text{ with}$$
$$xf^C = xg^C = a \quad \text{for each } x \in C_M .$$

In the terminology of the general theory of partial algebras, investigated in Part I by P. Burmeister, B is a subalgebra of C and A is a relative subalgebra of C.

The situation is quite different if we deal with the ep-signature Θ_{CAT} corresponding to small categories:

```
ΘCAT  is  sorts  Obj, Mor
      oprn   id(Obj) ──→ Mor
             source(Mor) ──→ Obj
             target(Mor) ──→ Obj
             comp(x:Mor,y:Mor iff x target = y source) ──→ Mor
      end  ΘCAT
```

In this case one can prove that any injective homomorphism $f: A \longrightarrow B$ between Θ_{CAT}-algebras reflects solutions of the domain condition def comp, so that any injective homomorphism is a closed homomorphism (see 2.2.10). Additionally, this implies that any bijective homomorphism is an isomorphism.

The preceding examples make it clear that we have to enrich the concept of an ep-signature if we want to exclude all the problems mentioned above.

If we look for a property of the ep-signature Θ_{CAT} which makes it qualitatively different from Θ_0 and Θ_1, we notice that in the ep-signatures Θ_0 and Θ_1 the operators are immediately or indirectly used in their own domain conditions, whereas in Θ_{CAT} some kind of hierarchy is given.

The aim of the following section is the investigation of this hierarchy condition.

2.4 Hierarchy Conditions

First of all we will give the concept of a hierarchical ep-signature a precise meaning. It must be kept clear in the sequel what it means that no operator is used immediately or indirectly in its own domain condition. For this reason we use the set op t, op(v:G) of operators which are syntactically used in any term $t \in T(\Sigma, v)$ and in any set of existence equations respectively, as defined at the end of Section 2.2.

Applying these definitions to the ep-signatures Θ_o, Θ_1, and Θ_{CAT} one gets

$$\text{op}(w: \{12...n\sigma = 12...n\sigma\}) = \{\sigma\}$$
$$\text{op}(\text{def } f) = \text{op}(w:\{x\} \longrightarrow \{M\} : \{xg = k\}) = \{g, k\}$$
$$\text{op}(\text{def comp}) = \{\text{source, target}\}$$
$$\text{op}(\text{def id}) = \emptyset .$$

<u>Definition 2.4.1</u>. An ep-signature $\Theta = (\Sigma, (\text{def } \sigma \mid \sigma \in \Omega))$ is said to be h i e r a r c h i c a l if every sequence of not necessarily pairwise distinct operators

$$\sigma_o, \sigma_1, \sigma_2, \ldots$$

with $\sigma_{i+1} \in \text{op}(\text{def } \sigma_i)$ for every $i \in \mathbb{N}$ terminates, or in other words, there is a natural number n with $\text{op}(\text{def } \sigma_n) = \emptyset$.
Hierarchical ep-signatures will be called h e p - s i g n a t u r e s for short. ■

Clearly, for the ep-signature Θ_o we can build an infinite sequence for every operator $\sigma \in \Omega_{\Theta_o}$, namely $\sigma, \sigma, \sigma, \ldots$ because $\sigma \in \text{op}(\text{def } \sigma)$, so that Θ_o is not a hierarchical ep-signature.
For the ep-signature Θ_1 we can construct the infinite sequence

$$f, g, f, g, \ldots, f, g, \ldots$$

because $f \in \text{op}(\text{def } g)$ and $g \in \text{op}(\text{def } f)$, so that Θ_1 is also not hierarchical.

As mentioned above Θ_{CAT} is hierarchical since any appropriate sequence of operators has at most the length two for $\text{op}(\text{def id}) = \text{op}(\text{def source}) = \text{op}(\text{def target}) = \emptyset$ and $\text{op}(\text{def comp}) = \{\text{source, target}\}$.

The importance of the concept of hep-signatures becomes evident by virtue of the following theorem.

<u>Theorem 2.4.2</u>. For any ep-signature $\Theta = (\Sigma, (\mathrm{def}\,\sigma \mid \sigma \in \Omega))$ and Θ-algebras A, B the following conditions are equivalent:

(i) Θ is hierarchical;

(ii) Every bijective homomorphism $f\colon A \longrightarrow B$ is an isomorphism;

(iii) Every injective homomorphism $f\colon A \longrightarrow B$ reflects solutions of existence equations;

(iv) Every injective homomorphism $f\colon A \longrightarrow B$ reflects solutions of domain conditions.

Before proving Theorem 2.4.2 we will make more precise the notion that a homomorphism $f\colon A \longrightarrow B$ reflects solutions of existence equations and of sets of existence equations respectively.

Let $(v\colon t_1 \overset{e}{=} t_2)$ and $(v{:}G)$ be any existence equation and any set of existence equations respectively. A homomorphism $f\colon A \longrightarrow B$ reflects solutions of $(v\colon t_1 \overset{e}{=} t_2)$ or of $(v{:}G)$ if for every assignment $\underline{a} \in A_v$

$$\underline{a}\, f_v \in B_{(v:\, t_1 \overset{e}{=} t_2)} \quad \text{implies} \quad \underline{a} \in A_{(v:\, t_1 \overset{e}{=} t_2)} \qquad \text{or}$$

$$\underline{a}\, f_v \in B_{(v:G)} \quad \text{implies} \quad \underline{a} \in A_{(v:G)}\,.$$

Proof of Theorem 2.4.2:

First we prove that (iv) implies (ii). Therefore let $f\colon A \longrightarrow B$ be a bijective homomorphism. Then there exists the inverse S-mapping

$$f^{-1} = (\, f_s^{-1}\colon B_s \longrightarrow A_s \mid s \in S\,).$$

We have to show that f^{-1} is a homomorphism. Let be $\sigma \in \Omega$, $\sigma\colon w \longrightarrow s$, $\underline{b} \in B_{\mathrm{def}\,\sigma}$ and $\underline{a} = \underline{b}\, f_w^{-1} \in A_w$. Because of $\underline{a}\, f_w = \underline{b} \in B_{\mathrm{def}\,\sigma}$ it follows according to (iv) that $\underline{a} = \underline{b}\, f_w^{-1} \in A_{\mathrm{def}\,\sigma}$. The injectivity of the homomorphism $f\colon A \longrightarrow B$ yields

$$\underline{b}\,\sigma^B f_s^{-1} = \underline{b}\, f_w^{-1} \sigma^A \quad \text{because}$$

$$\underline{b}\,\sigma^B f_s^{-1} f_s = \underline{b}\,\sigma^B = \underline{a}\, f_w \sigma^B = \underline{a}\,\sigma^A f_s = \underline{b}\, f_w^{-1}\sigma^A f_s\,.$$

This proves that $f^{-1}\colon B \longrightarrow A$ is a homomorphism.

Second we show indirectly that (i) is a consequence of (ii). For that purpose we define $\Omega_k \subseteq \Omega$ as the set of all those operators $\sigma \in \Omega$ for which there is a non-terminating sequence

$$\sigma = \sigma_o, \sigma_1, \ldots, \sigma_n, \ldots$$

with $\sigma_{i+1} \in op(def\,\sigma_i)$ for $i = 0,1,2,\ldots$.

If we assume that Θ is not hierarchical, then $\Omega_k = \emptyset$.
Analogously to the Θ_1-algebras in Section 2.3 we define two Θ-algebras F and F^* with $F^*_s = F_s = \{\emptyset\}$ for every $s \in S$. F is the uniquely determined total Θ-algebra, e.g.

$$dom\,\sigma^F = F_w \quad \text{for every } \sigma \in \Omega \text{ with } \sigma : w \longrightarrow s.$$

The fundamental operations of F^* are given by

$$dom\,\sigma^{F^*} = \begin{cases} F^*_w & \text{for every } \sigma \in \Omega \setminus \Omega_k, \text{ where } \sigma : w \longrightarrow s \\ \emptyset & \text{for every } \sigma \in \Omega_k. \end{cases}$$

The S-function

$$e = (e_s : F^* \longrightarrow F_s \mid s \in S)$$

with $e_s = Id_{\{\emptyset\}}$ for every $s \in S$ is a bijective homomorphism $e: F^* \longrightarrow F$, however it is not an isomorphism. This contradiction to (ii) can only be resolved if $\Omega_k = \emptyset$, which means that Θ is hierarchical.

It is not hard to show that (iv) holds provided (iii) is true. If $def\,\sigma = (w:\emptyset)$, where $\sigma : w \longrightarrow s$, then $B_{def\,\sigma} = B_w$ for every Θ-algebra, so that for $\underline{a} \in A_w$ with $\underline{a}\,f_w \in B_{def\,\sigma}$ obviously $\underline{a} \in A_{def\,\sigma} = A_w$. Now we assume $def\,\sigma \neq (w:\emptyset)$, $\underline{a} \in A_w$ and $\underline{a}\,f_w \in B_{(w:\,t_1 \overset{e}{=} t_2)}$ for every $(w:\,t_1 \overset{e}{=} t_2) \in def\,\sigma$, so that by (iii) $\underline{a} \in A_{(w:\,t_1 \overset{e}{=} t_2)}$ for every $(w:\,t_1 \overset{e}{=} t_2) \in def\,\sigma$. This means $\underline{a} \in A_{def\,\sigma}$ which is what we have to show.

Next we turn to the implication (iv) $\Longrightarrow$ (iii). By a simple induction one can prove that (iv) implies the following slightly generalized condition

(iv)* Every injective homomorphism $f: A \longrightarrow B$ reflects solutions of $def\,t$ for any term $t \in T(\Sigma,u)_s$, $u: Y \longrightarrow S$.

The proof goes like the proof of Corollary 2.3.5. Let $(v:\,t_1 \overset{e}{=} t_2)$ with $v: X \longrightarrow S$ be any existence equation, $\underline{a} \in A_v$ and $f: A \longrightarrow B$ any injective homomorphism with

$$\underline{a}\,f_v \in B_{(v:\,t_1 \overset{e}{=} t_2)}.$$

This implies $\underline{a}\,f_v \in B_{def\,t_1}$, $\underline{a}\,f_v \in B_{def\,t_2}$, and $\underline{a}\,f_v\,t_1^B = \underline{a}\,f_v\,t_2^B$.

Condition (iv)* together with the definition of the term functions t_1^B, t_2^B yield

$$\underline{a} \in A_{\mathrm{def}\, t_1},\ \underline{a} \in A_{\mathrm{def}\, t_2}, \text{ and } t_1(\widetilde{\underline{a} f_v}) = t_2(\widetilde{\underline{a} f_v}) .$$

Since $f: A \longrightarrow B$ is a homomorphism we obtain by (2.3.4)

$$t_1 \underline{\widetilde{a}}_s f_s = t_1(\widetilde{\underline{a} f_v}) = t_2(\widetilde{\underline{a} f_v}) = t_2 \underline{\widetilde{a}}_s f_s ,$$

where $s \in S$ is the sort name with $t_1, t_2 \in T(\Sigma, v)_s$.

The injectivity of $f_s: A_s \longrightarrow B_s$ yields

$$t_1 \underline{\widetilde{a}} = t_2 \underline{\widetilde{a}} , \text{ so that } \underline{a}\, t_1^A = \underline{a}\, t_2^A \text{ or in other words}$$

$$\underline{a} \in A_{(v:\, t_1 \overset{e}{=} t_2)} .$$

We conclude the proof of Theorem 2.4.2 by showing that (i) implies (iv) . This is the main part of the proof. The presupposed hierarchy of Θ allows us to prove (iv) by the following principle of induction: Let Ω be a poset (partially ordered set) in which every descending chain is finite. Let P be a subset of Ω such that P contains all minimal elements of Ω and such that for every $\sigma \in \Omega$ if $\{ \sigma' \in \Omega \mid \sigma' \leq \sigma ,\ \sigma' \neq \sigma \} \subseteq P$ then $\sigma \in P$. Under these conditions $P = \Omega$.

First we show that (iv) holds for all operators $\sigma \in \Omega$ with $op(\mathrm{def}\,\sigma) = \emptyset$.

Since (iv) clearly holds for all $\sigma \in \Omega$ with $\sigma: w \longrightarrow s$ and $\mathrm{def}\,\sigma = (w: \emptyset)$, condition (iv) also holds for every $\sigma \in \Omega$ with $op(\mathrm{def}\,\sigma) = \emptyset$ and $\sigma: \lambda \longrightarrow s$. If $\sigma: w \longrightarrow s$, $w: [n] \longrightarrow S$, $n \neq 0$, and $op(\mathrm{def}\,\sigma) = \emptyset$ then

$$\mathrm{def}\,\sigma = (w: \{ i_1 \overset{e}{=} j_1 , \ldots , i_m \overset{e}{=} j_m \})$$

with $i_k, j_k \in [n]$ for every $k = 1,2,\ldots,m$.

Let be $\underline{a} \in A_w$ and $f: A \longrightarrow B$ an injective homomorphism with

$$\underline{a} f_w \in B_{(w: \{ i_1 \overset{e}{=} j_1 , \ldots , i_m \overset{e}{=} j_m \})} .$$

Since each $(w: i_k \overset{e}{=} j_k)$ for $k \in [m]$ is an existence equation, it follows that $i_k w = j_k w$ for each $k \in [m]$.

Now $\underline{a} f_w \in B_{\mathrm{def}\,\sigma}$ implies $(\underline{a} f_w) i_k^B = (\underline{a} f_w) j_k^B$ for each $k \in [m]$.

Consequently

$$i_k(\underline{a} f_w) = j_k(\underline{a} f_w) \text{ or } (i_k \underline{a}) f_{i_k w} = (j_k \underline{a}) f_{j_k w} \text{ for } k \in [m] ,$$

which implies together with the injectivity of f that $i_k \underline{a} = j_k \underline{a}$ for every $k \in [m]$, and therefore $\underline{a} \in A_{\mathrm{def}\,\sigma}$.

So far we have proved that (iv) is true for every operator $\sigma \in \Omega$ with $op(\mathrm{def}\,\sigma) = \emptyset$. For the proof of this condition (i) was not used.

For $\sigma \in \Omega$ with $op(\mathrm{def}\,\sigma) = \emptyset$ we set

$$P(\sigma) = \{\sigma' \in \Omega \mid \sigma' \neq \sigma \text{ and there is a finite sequence } \sigma = \sigma_1, \sigma_2, \ldots, \sigma_n = \sigma' \text{ with } \sigma_{i+1} \in op(\mathrm{def}\,\sigma_i) \text{ for } i=1,2,\ldots,n-1\}.$$

Now we assume that (iv) is true for all $\sigma' \in P(\sigma)$, so that we have to prove that (iv) holds also for σ. Let $\sigma \in \Omega$, $\sigma: w \longrightarrow s$, $f: A \longrightarrow B$ any injective homomorphism, $\underline{a} \in A_w$ with $\underline{a} f_w \in A_{\mathrm{def}\,\sigma}$, and let $(w: t_1 \overset{e}{=} t_2) \in \mathrm{def}\,\sigma$ be any existence equation of $\mathrm{def}\,\sigma$.

Then $op(t_1) \subseteq P(\sigma)$ and $op(t_2) \subseteq P(\sigma)$. A simple induction shows that (iv)* holds for every term $t \in T(\Sigma, v)$, $v: X \longrightarrow S$, with $op(t) \subseteq P(\sigma)$ provided (iv) holds for every $\sigma' \in P(\sigma)$.

Now $\underline{a} f_w \in B_{\mathrm{def}\,\sigma}$ implies $\underline{a} f_w \in B_{(w:\, t_1 \overset{e}{=} t_2)} \supseteq B_{\mathrm{def}\,\sigma}$.

This means that

$$\underline{a} f_w \in B_{\mathrm{def}\, t_1}, \quad \underline{a} f_w \in B_{\mathrm{def}\, t_2}, \text{ and } \underline{a} f_w t_1^B = \underline{a} f_w t_2^B .$$

Therefore $\underline{a} \in A_{\mathrm{def}\, t_1}$, $\underline{a} \in A_{\mathrm{def}\, t_2}$ and analogously to the proof of of (iv) $\Longrightarrow$ (iii) one can see that

$$\underline{a} \in A_{(w:\, t_1 \overset{e}{=} t_2)} \quad \text{for every } (w: t_1 \overset{e}{=} t_2) \in \mathrm{def}\,\sigma .$$

This means that $\underline{a} \in A_{\mathrm{def}\,\sigma}$. In this manner we see that (iv) holds for every $\sigma \in \Omega$.

This concludes the proof of Theorem 2.4.2.

Theorem 2.4.2 makes apparent the great importance of the hierarchy condition for a proper theory of partial algebras. We will see that hierarchical equationally partial algebras not only enable a proper theory but that this concept is general enough to treat a wide spectrum of applications in algebra as well as in computer science.

We call Θ-algebras for hierarchical ep-signatures

E Q U O I D s

for short, which is an artificial name.

Definition 2.4.3. Let $\Theta = (\Sigma, (\text{def}\, \sigma \mid \sigma \in \Omega))$ be any hep-signature. A Θ-equoid A is said to be a s u b e q u o i d of the Θ-equoid B if

(i) $A_s \subseteq B_s$ for each $s \in S_\Theta$ and

(ii) the S-inclusion $i = (i_s : A_s \longrightarrow B_s \mid s \in S_\Theta)$ with $x\, i_s = x$ for each $s \in S_\Theta$ and $x \in A_s$ forms a homomorphism. ∎

Just as in the case of total algebras we can characterize those S-subsets of an equoid A which are carriers of a subequoid as those which are closed with respect to all fundamental operations σ^A.

Corollary 2.4.4. Let $\Theta = (\Sigma, (\text{def}\, \sigma \mid \sigma \in \Omega))$ be any hep-signature and $A \in ALG(\Theta)$ any equoid. An S-subset $M \subseteq A$ is a carrier of a subequoid of A if and only if for every $\sigma \in \Omega$ with $\sigma : w \longrightarrow s$ and every $\underline{m} \in M_w \cap A_{\text{def}\,\sigma}$ it follows that

$$\underline{m}\, \sigma^A \in M_s .$$

Proof. If M is a subequoid of A, then the S-inclusion $i : M \longrightarrow A$ is a homomorphism, so that for any $\sigma \in \Omega$ with $\sigma : w \longrightarrow s$, and any $\underline{m} \in M_w \cap A_{\text{def}\,\sigma}$ it follows that $\underline{m}\, i_w \in A_{\text{def}\,\sigma}$. As a consequence of Theorem 2.4.2 we obtain

$$\underline{m} \in M_{\text{def}\,\sigma}, \text{ and } \underline{m}\, \sigma^A = \underline{m}\, i_w\, \sigma^A = \underline{m}\, \sigma^M f_s = \underline{m}\, \sigma^M \in M_s .$$

Secondly, we assume that M is closed with respect to all fundamental operations of A. For every $\sigma \in \Omega$ with $\sigma : w \longrightarrow s$ we define a partial operation

$$\sigma^M : \text{dom}\, \sigma^M \longrightarrow M_s \text{ by } \text{dom}\, \sigma^M = M_w \cap A_{\text{def}\,\sigma} \text{ and } \underline{m}\, \sigma^M = \underline{m}\, \sigma^A$$

for every $\underline{m} \in M_w \cap A_{\text{def}\,\sigma}$.

Thus we have to show that

$$M = ((M_s \mid s \in S), (\sigma^M \mid \sigma \in \Omega))$$

is not only a partial algebra of type $(S, \alpha : \Omega \longrightarrow S^* \times S)$ but also an equoid of type Θ.

One can easily prove inductively that for any S-sorted system of variables $v : X \longrightarrow S$ the following holds:

(i) If $t \in T(\Sigma, v)_s$ and $\underline{m} \in \text{dom}\, t^A \cap M_v$ then $\underline{m}\, t^A \in M_s$.

(ii) $\text{dom}\, t^M = \text{dom}\, t^M \cap M_v$

(iii) If we consider the term function $t^M : \text{dom}\, t^M \longrightarrow M_s$ as a subset of $M_v \times M_s$ with $(\underline{m}, m) \in t^M$ iff $m = \underline{m}\, t^{M^s}$ then

$$t^M = t^A \cap (M_v \times M_s) .$$

Similarly to the proof of Theorem 2.4.2 we can continue here and show by induction on the hierarchy of operators that

$$\mathrm{dom}\,\sigma^M = M_{\mathrm{def}\,\sigma} \ .$$

Therefore as before we set

$$P(\sigma) = \{\sigma' \in \Omega \mid \text{there is a finite sequence } \sigma = \sigma_1, \ldots, \sigma_n = \sigma' \text{ with } \sigma_{i+1} \in \mathrm{op}(\mathrm{def}\,\sigma_i) \text{ for } i = 1,2,\ldots,n-1\} .$$

If $P(\sigma) = \emptyset$ then either $\mathrm{def}\,\sigma = (w{:}\emptyset)$ or

$$\mathrm{def}\,\sigma = (w{:}\{i_1 \stackrel{e}{=} j_1, \ldots, i_r \stackrel{e}{=} j_r\})$$

where $i_k, j_k \in [n]$ for every $k = 1,2,\ldots,r$ and where $\sigma : w \longrightarrow s$ and $w{:}\,[n] \longrightarrow S$.

$\mathrm{def}\,\sigma = (w{:}\emptyset)$ implies $M_{(w:\emptyset)} = M_w = M_w \cap A_w = M_w \wedge A_{(w:\emptyset)} =$ $= M_w \cap \mathrm{dom}\,\sigma^A = \mathrm{dom}\,\sigma^M$, and in the other case

$$M_{\mathrm{def}\,\sigma} = \{\underline{m} \in M_w \mid i_k\underline{m} = j_k\underline{m} \text{ for every } k = 1,2,\ldots,r\} =$$
$$= M_w \cap \{\underline{m} \in A_w \mid i_k\underline{m} = j_k\underline{m} \text{ for every } k=1,2,\ldots,r\} =$$
$$= M_w \cap A_{\mathrm{def}\,\sigma} = \mathrm{dom}\,\sigma^M \ .$$

Now we turn to the case $P(\sigma) \neq \emptyset$ and show that $M_{\mathrm{def}\,\sigma} = \mathrm{dom}\,\sigma^M$ holds provided $M_{\mathrm{def}\,\sigma'} = \mathrm{dom}\,(\sigma')^M$ holds for every $\sigma' \in P(\sigma)$. This assumption leads to $M_{\mathrm{def}\,t} = \mathrm{dom}\,t^M$ for every $v{:}\,X \longrightarrow S$, $s \in S$, $t \in T(\Sigma, v)_s$ with $\mathrm{op}(t) \subseteq P(\sigma)$, so we obtain

$$\mathrm{dom}\,\sigma^M = \mathrm{dom}\,\sigma^A \cap M_w = A_{\mathrm{def}\,\sigma} \cap M_w =$$
$$= \bigcap \{A_{(w:\ t_1 \stackrel{e}{=} t_2)} \mid (w{:}\ t_1 \stackrel{e}{=} t_2) \in \mathrm{def}\,\sigma\} \cap M_w =$$
$$= \bigcap (\{A_{\mathrm{def}\,t_1} \cap A_{\mathrm{def}\,t_2} \mid (w{:}\ t_1 \stackrel{e}{=} t_2) \in \mathrm{def}\,\sigma\} \cap$$
$$\cap \{\underline{a} \in A_w \mid \underline{a}\,t_1^A = \underline{a}\,t_2^A \ , (w{:}\ t_1 \stackrel{e}{=} t_2) \in \mathrm{def}\,\sigma\}) \wedge M_w =$$
$$= \{\underline{m} \in M_w \mid \underline{m} \in \mathrm{dom}\,t_1^M \cap \mathrm{dom}\,t_2^M,\ \underline{m}\,t_1^M = \underline{m}\,t_2^M,$$
$$(w{:}\ t_1 \stackrel{e}{=} t_2) \in \mathrm{def}\,\sigma\} =$$
$$= M_{\mathrm{def}\,\sigma} \ . \qquad \blacksquare$$

The problem we now deal with is the behaviour of subequoids under homomorphisms.

An essential difference from the behaviour of total algebras appears now. Homomorphisms between equoids in general do not preserve subequoids, e.g. the image of a subequoid is in general not the carrier of a subequoid. This fact is very well known in the theory of small categories. We illustrate this difference by the following homomorphism $f: A \longrightarrow B$ between Θ_{CAT}-equoids A and B.

Let A be the Θ_{CAT}-equoid with carrier sets

$$A_{Obj} = \{1,2,3,4\}, \quad A_{Mor} = \{e_1,e_2,e_3,e_4,x,y\},$$

and with fundamental operations given by

$$j\, id^A = e_j \quad \text{for} \quad j = 1,2,3,4$$

$$x\, source^A = e_1\, source^A = e_1\, target^A = 1$$

$$x\, target^A = e_2\, source^A = e_2\, target^A = 2$$

$$y\, source^A = e_3\, source^A = e_3\, target^A = 3$$

$$y\, target^A = e_4\, source^A = e_4\, target^A = 4$$

$$dom\, comp^A = \{(e_1,x),(x,e_2),(e_3,y),(y,e_4),(e_1,e_1),(e_2,e_2),(e_3,e_3),(e_4,e_4)\}$$

$$comp^A(e_j,e_j) = e_j \quad \text{for} \quad j = 1,2,3,4$$

$$comp^A(e_1,x) = comp^A(x,e_2) = x$$

$$comp^A(e_3,y) = comp^A(y,e_4) = y .$$

Let B be the Θ_{CAT}-equoid with carrier sets

$$B_{Obj} = \{I,II,III\}, \quad B_{Mor} = \{E_1,E_2,E_3,X,Y,Z\},$$

and with fundamental operations given by

$$I\, Id^B = E_1,\ II\, Id^B = E_2,\ III\, Id^B = E_3,$$

$$X\, source^B = Z\, source^B = E_1\, source^B = E_2\, target^B = I,$$

$$X\, target^B = E_2 source^B = E_2\, target^B = Y\, source^B = II,$$

$$Y\, target^B = Z\, target^B = E_3\, source^B = E_3\, target^B = III,$$

$$dom\, comp^B = \{(E_1,X),(X,E_2),(E_2,Y),(Y,E_3),(X,Y),(E_1,Z),(Z,E_3),(E_1,E_1),(E_2,E_2),(E_3,E_3)\}.$$

$$comp^B(E_j,E_j) = E_j \quad \text{for} \quad j = 1,2,3,$$

$$comp^B(E_1,X) = comp^B(X,E_2) = X,$$

$$comp^B(E_2,Y) = comp^B(Y,E_3) = Y,$$

$$comp^B(E_1,Z) = comp^B(Z,E_3) = comp^B(X,Y) = Z .$$

Let $f: A \longrightarrow B$ be the homomorphism defined by

$1 f_{Obj} = I$, $2 f_{Obj} = 3 f_{Obj} = II$, $4 f_{Obj} = III$,

$e_1 f_{Mor} = E_1$, $e_2 f_{Mor} = e_3 f_{Mor} = E_2$, $e_4 f_{Mor} = E_3$,

$x f_{Mor} = X$, $y f_{Mor} = Y$.

The image $A f$ B is not the carrier of a subequoid of B, since

$$(x f_{Mor}, y f_{Mor}) \in \mathrm{dom}\, \mathrm{comp}^B \quad \text{and}$$

$$Z = \mathrm{comp}^B(x f_{Mor}, y f_{Mor}) \notin A f_{Mor}.$$

Certainly we can find a simpler counter-example, but as a side effect we want to demonstrate how laborious it is at this underdeveloped state of the theory to describe concrete equoids. Therefore we will extend the well-known practice of defining groups by generators and relations to equoids. The formalism for doing this is the main point of Chapter 3.

It was not possible to prove that homomorphisms preserve subequoids; however, we can show that homomorphisms reflect subequoids.

Corollary 2.4.5. Let Θ be any hep-signature, $f: A \longrightarrow B$ any Θ-homomorphism and $C \subseteq B$ a subequoid of B. Then the inverse image

$$C f^{-1} = (C_s f_s^{-1} \mid s \in S) \text{ where } C_s f_s^{-1} = \{x \in A_s \mid x f_s \in C_s\}$$

for every $s \in S$, is the carrier of a subequoid of A.

Proof. In view of the preceding corollary we have only to show that $C f^{-1}$ is closed. To this end let be given $\sigma \in \Omega$ with $\sigma: w \longrightarrow s$ and $\underline{a} \in (C f^{-1})_w \cap \mathrm{dom}\, \sigma^A$. Since every homomorphism preserves solutions of domain conditions and since C is a subequoid it follows that

$$\underline{a} f_w \in \mathrm{dom}\, \sigma^B, \text{ so that } \underline{a} f_w \sigma^B = \underline{a} \sigma^A f_s \in C_s$$

or in other words

$$\underline{a} \sigma^A \in C_s f_s^{-1}. \qquad \blacksquare$$

From (2.4.5) it is easy to show that subequoids are closed under intersection. Thus, we can define the notion of the subequoid generated by an S-subset.

Definition 2.4.6. Let Θ be any hep-signature, A any Θ-equoid, $X \subseteq A$ any S-subset of the carrier of A, and $f: A \longrightarrow B$ any homomorphism between Θ-equoids.

By $\langle X \rangle$ we denote the smallest subequoid of A containing X. $\langle X \rangle$ is

said to be the s u b e q u o i d of A g e n e r a t e d b y X .
In connection with homomorphisms $\langle A f\rangle$ will be referred to as the h o m o m o r p h i c i m a g e of A. ∎

An immediate consequence of Corollary 2.4.4 and Definition 2.4.6 is the following characterization of the elements of X

Corollary 2.4.7. Let Θ be any hep-signature, A any Θ-equoid and $X \subseteq A$ any S-subset. Then $a \in \langle X\rangle_s$, $s \in S$, if and only if there are $w:[n] \longrightarrow S$, $t \in T(\Sigma, w)_s$, $\underline{x} \in X_w$ with $\underline{x} \in \operatorname{dom} t^A$ and $a = \underline{x}\, t^A$.

Proof. Obviously, the S-set of all those elements of A which satisfy the condition of (2.4.7) contains X and is closed with respect to all fundamental operations of A , so that this S-subset is the carrier of a subequoid containing X . Hence, every element $a \in \langle X\rangle$ satisfies this condition.
By a simple induction one can generalize (2.4.4) in such a way that any subequoid is closed not only with respect to fundamental operations but also with respect to term functions. ∎

For the homomorphic image $\langle A f\rangle$ of a homomorphism $f: A \longrightarrow B$ one has to pay attention to the fact that the S-subset $A f$ in general is not the carrier of $\langle A f\rangle$ but it is only a generating set.

A further important algebraic construction is the construction of quotient algebras, also called factor algebras, by identifying elements. Usually for total algebras this construction is performed by so-called congruence relations, i.e. equivalence relations which are closed with respect to fundamental operations. In the basic theory of partial algebras the concept of a quotient algebra cannot be generalized in a straightforward way. We will not use an element-wise construction of quotient algebras but rather the universal property of the resulting natural homomorphism. At this point we will give only a sketch of the problem, since it will be extensively discussed in Section 3.3.

Definition 2.4.8. Let Θ be any hep-signature, A , Q any Θ-equoids and ϱ any binary S-relation in the carrier of A .
A homomorphism $q: A \longrightarrow Q$ is said to be n a t u r a l for ϱ if:

(1) For every $s \in S$ and every $(x,y) \in \varrho_s$, $x q_s = y q_s$ holds .

(2) If $f: A \longrightarrow B$ is a homomorphism between Θ-equoids A and B such that $x f_s = y f_s$ for every $s \in S$ and every $(x,y) \in \varrho_s$, then

there exists exactly one homomorphism $h: Q \longrightarrow B$ with $f = q \cdot h$.
If $f: A \longrightarrow Q$ is natural for an S-relation ϱ in A, then the equoid Q is called a q u o t i e n t e q u o i d of A with respect to ϱ. ∎

If for an S-relation ϱ in the carrier of an equoid A natural homomorphisms $q_1: A \longrightarrow Q_1$, $q_2: A \longrightarrow Q_2$ exist, then there must be an isomorphism $h: Q_1 \longrightarrow Q_2$ with $q_2 = q_1 \cdot h$, so natural homomorphisms are unique up to isomorphism.

This notion of a natural homomorphism is derived from categorical structure theory, see MacLane (1971), and it coincides with the notion of the natural homomorphism

$$\text{nat}\,\varrho : A \longrightarrow A/\varrho$$

of a congruence relation in a total algebra associating with every element $a \in A$ its congruence class $[a]_\varrho = a(\text{nat}\,\varrho)$.

Here we mention only that a natural homomorphism between equoids is in general not a surjective S-function. The problems of existence and of construction of quotient equoids will be treated in Section 3.3. The problems arise from the fact that by set-theoretic factorizations new solutions of domain conditions may be created in such a way that the value of the fundamental operation for this new solution cannot be reduced to any representative of a congruence class. To illustrate this we return to the Θ_{CAT}-equoid A as defined above. If we choose the S-relation ϱ given by

$$\varrho_{Obj} = \{(2,3)\}, \quad \varrho_{Mor} = \{(e_2,e_3)\}$$

then what we just described happens.

2.5 hep-Varieties and hep-Quasi-Varieties

Until now we have only generalized the concept of a signature, and we have not yet any means available to describe such properties of operators of sets of operators as associativity, commutativity, or distributivity.

As mentioned in Section 2.2 no kind of validity of term equations is known which is sufficient for describing the basic properties of the set of operations for small categories. Therefore we need other means, which have not a lower expressiveness than existence equations have. Just as for signatures we use the theory of small categories as a guide to discover suitable means. Let us consider the associativity law of small categories, which says that for any morphisms x,y,z the equation

$$comp^A(comp^A(x,y),z) = comp^A(x,comp^A(y,z))$$

holds, provided $comp^A(x,y)$, $comp^A(y,z)$, $comp^A(comp^A(x,y),z)$, and $comp^A(x,comp^A(y,z))$ are all defined. This kind of conditional equation can be formalized in the following way.

Definition 2.5.1. Let $\Sigma = (S, \alpha : \Omega \longrightarrow S^* \times S)$ be any signature. An elementary implication

$$(v\colon G \longrightarrow t_1 \overset{e}{=} t_2)$$

is given by a finite S-sorted system of variables $v\colon X \longrightarrow S$, the domain of the elementary implication, by an existence equation $(v\colon t_1 \overset{e}{=} t_2)$, the conclusion of the elementary implication, and by a finite set of existence equations $(v\colon G)$, the premise of the elementary implication.
The elementary implication $(v\colon G \longrightarrow t_1 \overset{e}{=} t_2)$ holds in a partial Σ-algebra A if

$$A_{(v\colon G)} \subseteq A_{(v\colon\, t_1 \overset{e}{=} t_2)} .$$

This will be denoted briefly by

$$A \models (v\colon G \longrightarrow t_1 \overset{e}{=} t_2) .$$

An implication

$$(v\colon G \longrightarrow H)$$

is given by an S-sorted system of variables $v\colon X \longrightarrow S$ and by two sets of existence equations $(v\colon G)$, $(v\colon H)$.
$(v\colon G \longrightarrow H)$ holds in a partial Σ-algebra A, denoted by

$$A \models (v\colon G \longrightarrow H) \text{ , if } A_{(v:G)} \subseteq A_{(v:H)} .$$ ∎

We draw attention to the fact that implications can cause demands of applicability. We illustrate this by means of the hep-signature Θ_{CAT} to which we add the elementary implication

$$(w\colon \emptyset \longrightarrow comp(comp(1,2),3) \overset{e}{=} comp(1,comp(2,3)))$$

where $w\colon[3] \longrightarrow \{Obj,Mor\}$ with $1w = 2w = 3w = Mor$.
If a Θ_{CAT}-equoid A satisfies this elementary implication, then

$$A_{(w:\emptyset)} = A_w \subseteq A_{(w:comp(comp(1,2),3) \overset{e}{=} comp(1,comp(2,3)))}$$

holds, so that every assignment $\underline{a} \in A_w$ is a solution of the existence equation $(w: comp(comp(1,2),3) \overset{e}{=} comp(1,comp(2,3)))$. This means that $comp^A(x,y)$ is defined for every $(x,y) \in A_{Mor} \times A_{Mor}$. Thus, $comp^A$ is a total binary multiplication. Consequently, we have used the wrong implication to give the associativity law of small categories.

The class of small categories is given by those Θ_{CAT}-equoids satisfying the following set $\mathcal{O}_{CAT}$ of elementary implications:

$$
\begin{aligned}
\mathcal{O}_{CAT} = \; & (v: \emptyset \longrightarrow source(id(e)) \overset{e}{=} e), (v: \emptyset \longrightarrow target(id(e)) \overset{e}{=} e), \\
& (v: \emptyset \longrightarrow comp(id(source(x)),x) = x), \\
& (v: \{target(x) = source(y)\} \longrightarrow source(x)=sourc(comp(x,y))), \\
& (v: \{target(x)=source(y)\} \longrightarrow target(y)=target(comp(x,y))), \\
& (v: \{target(x)=source(y), target(y)=source(z)\} \\
& \quad \longrightarrow comp(comp(x,y),z)=comp(x,comp(y,z)))
\end{aligned}
$$

where $v: \{e,x,y,z\} \longrightarrow \{Obj,Mor\}$ with $e\,v = Obj$ and $x\,v = y\,v = z\,v = Mor$.

The total unary operators $source(Mor) \longrightarrow Obj$, $target(Mor) \longrightarrow Obj$ associate with every morphism $x \in A_{Mor}$ its so-called source object $source^A(x) \in A_{Obj}$, and target object $target^A(x) \in A_{Obj}$. The unary total operator $id(Obj) \longrightarrow Mor$ associates with every object $e \in A_{Obj}$ its so-called identity morphism $id^A(e) \in A_{Mor}$ which has, according to the first two elementary implications, the given object e as its source and target object. For any morphism $x \in A_{Mor}$ the morphisms $id^A(source^A(x)) \in A_{Mor}$ and $id^A(target^A(x)) \in A_{Mor}$ are called the left identity and right identity of x respectively. Their properties are given by the third and fourth elementary implications. The equationally partial binary operator

$$comp(x:Mor,y:Mor \; \underline{iff} \; target(x)=source(y)) \longrightarrow Mor$$

associates with any ordered pair of morphisms $(x,y) \in A_{Mor} \times A_{Mor}$ a morphism $comp^A(x,y) \in A_{Mor}$ provided the target object of x equals the source object of y , i.e. iff (x,y) is a solution of the domain condition. If $comp^A(x,y)$ is defined, then its source object equals the source object of the first factor, according to the fifth elementary implication, and target objects of $comp^A(x,y)$ and the second factor are equal, according to the sixth elementary implication. The last elementary implication is the correct form of the associativity law of the partial binary composition.

A typical example of a small category can be obtained if one takes $A_{Obj} = M$ any set of sets, A_{Mor} = the set of all functions $f: X \longrightarrow Y$

with $X,Y \in M$. The fundamental operations of the small category are then given by

$id^A(X) = Id_X: X \longrightarrow X$ for every $X \in M = A_{Obj}$,

$source^A(f: X \longrightarrow Y) = X$, $target^A(f: X \longrightarrow Y) = Y$ for every $(f: X \longrightarrow Y) \in A_{Mor}$, and

$comp^A$ is the composition of functions.

It is not the aim of this book to give an introduction to the theory of categories, although the author completely agrees with Burstall and Goguen who consider this theory as very useful for computer scientists, see Goguen, Burstall (1983).

As the example of small categories shows, interesting classes of equoids can be described by a hep-signature Θ with a set $\mathcal{A}$ of elementary implications acting as axioms.

But the class of small categories could also be described by an ordinary signature and a modified set of elementary implications, namely by Σ_{CAT} and $\mathcal{A}^*_{CAT}$ given by

Σ_{CAT} <u>is</u> <u>sorts</u> Obj,Mor
<u>oprn</u> id(Obj) ⟶ Mor
source(Mor) ⟶ Obj
target(Mor) ⟶ Obj
comp(Mor,Mor) —o⟶ Mor
<u>end</u>

$$\mathcal{A}^*_{CAT} = \mathcal{A}_{CAT} \cup \{(w: \{comp(x,y) \overset{e}{=} comp(x,y)\} \longrightarrow target(x) \overset{e}{=} source(y)), (w: \{target(x) \overset{e}{=} source(y)\} \longrightarrow comp(x,y) \overset{e}{=} comp(x,y))\}$$

The first additional axiom represents the necessity and the second axiom represents the sufficiency.

Thus, we have at least two formalisms for describing the classes of partial algebras. On the one hand we can use ordinary signatures and sets of elementary implications, and on the other hand we can use hep-signatures and sets of elementary implications.

At first glance the first approach seems to be simpler. However, in this case domain conditions and laws interrelating operators are mixed together and both are represented by elementary implications. Not every set of elementary implications is so well structured as $\mathcal{A}^*_{CAT}$, where we can see which implications give domain conditions and which ones describe operator interrelations.

The second approach is more structured, but not enough. Domain conditions may be expressed both by hep-signatures and by the set of axioms. Therefore we will restrict the set of axioms in such a way that no additional domain conditions are expressed by the axioms. Then we have a well-structured approach. But it is not a mathematical question which of the two approaches should be favoured. The choice is largly a matter of philosophical taste, together with practical needs.

We prefer the more structured approach where the single parts of the formalism are determined by their functions. Therefore, the hep-signatures are used to describe all required domain conditions and the set $\mathcal{A}$ of axioms serves to express laws interrelating operations only. Consequently, we are now going to restrict the set of axioms in the announced way.

As preparation we need the concept of c o n s e q u e n c e s of a given set of elementary implications. First, we denote the fact that a partial algebra A satisfies all axioms of $\mathcal{A}$ by

$$A \models \mathcal{A} .$$

An elementary implication $(v\colon G \longrightarrow t_1 \overset{e}{=} t_2)$ is said to be a consequence of the set $\mathcal{A}$ of elementary implications if for every partial Σ-algebra A the statement $A \models \mathcal{A}$ implies $A \models (v\colon G \longrightarrow t_1 \overset{e}{=} t_2)$. The fact that $\mathcal{A}$ entails $(v\colon G \longrightarrow t_1 \overset{e}{=} t_2)$ will be denoted by

$$\mathcal{A} \models (v\colon G \longrightarrow t_1 \overset{e}{=} t_2) .$$

In the same way the corresponding notions are defined for arbitrary implications $(v\colon G \longrightarrow H)$ and sets of arbitrary implications.

In Section 3.1 we will prove a completeness theorem, so that this semantic definition can be replaced by a syntactic one. In the meantime we can give one way of restricting the set of axioms.

<u>Definition 2.5.2</u>. Θ may be any ep-signature. A finite set $\mathcal{A}$ of elementary implications is said to be s t a b l e if there exists a partial order $\leq$ on $\mathcal{A}$ such that for every $(v\colon G \longrightarrow t_1 \overset{e}{=} t_2)$ the following condition is satisfied:

$(v\colon G \longrightarrow t_1 \overset{e}{=} t_2)$ is a consequence of

$$\{(u\colon H \longrightarrow r_1 \overset{e}{=} r_2) \in \mathcal{A} \mid (u\colon H \longrightarrow r_1 \overset{e}{=} r_2) \lneqq (v\colon G \longrightarrow t_1 \overset{e}{=} t_2)\} ,$$

i.e. it is a consequence of all those elementary implications which are

less than but not equal to the given one.
If such a partial order on $\mathcal{A}$ does not exist then $\mathcal{A}$ is called u n s t a b l e . ∎

The set $\mathcal{A}_{CAT}$ is an example of a stable set of axioms. To see this let us denote the single elementary implications of $\mathcal{A}_{CAT}$ by $a_1, a_2, \ldots, a_7$ respectively. We are going to derive a proper partial order on $\mathcal{A}_{CAT}$. Because def(v: source(id(e))$\stackrel{e}{=}$e) = (v: $\emptyset$) and def(v: target(id(e))$\stackrel{e}{=}$e) = (v: $\emptyset$) axioms a_1 and a_2 can be taken as minimal ones. If we consider axioms a_5 and a_6, then

def(v: source(x)$\stackrel{e}{=}$source(comp(x,y))) = (v: target(x)$\stackrel{e}{=}$source(y)) =
= def(v: target(y)$\stackrel{e}{=}$target(comp(x,y))) ,

so that a_5 and a_6 can also be taken as minimal ones since

$\emptyset \models$ (v: { target(x)$\stackrel{e}{=}$source(y) } ⟶ target(x)$\stackrel{e}{=}$source(y)).

Owing to the fact that

def(v: comp(id(source(x)))$\stackrel{e}{=}$x) = (v: target(id(source(x))) $\stackrel{e}{=}$ source(x))

the implication (v: $\emptyset$ ⟶ target(id(source(x))) $\stackrel{e}{=}$ source(x)) should be a consequence of a suitable set of minimal axioms. Indeed, this implication is a consequence of $\{a_2\}$ if we replace e by source(x) . In the same way one can see that

(v: $\emptyset$ ⟶ target(x) $\stackrel{e}{=}$ source(id(target(x))))

is a consequence of $\{a_1\}$, where

def(v: comp(x,id(target(x)))$\stackrel{e}{=}$x) = (v: target(x) $\stackrel{e}{=}$ source(id(target(x)))).

Finally

def(v: comp(x,comp(y,z))=comp(comp(x,y),z)) =
=(v: { target(x)$\stackrel{e}{=}$source(comp(x,y) , target(comp(x,y)) $\stackrel{e}{=}$ source(z)})

holds, so that

(v: { target(x)$\stackrel{e}{=}$source(y) , target(y)$\stackrel{e}{=}$source(z)} ⟶
⟶ {target(x)$\stackrel{e}{=}$source(comp(y,z)), target(comp(x,y))$\stackrel{e}{=}$source(z)})

has to be a consequence of the remaining axioms. It is not hard to see that this implication is a consequence of $\{a_5, a_6\}$. Therefore, the partial order is given by $a_1 < a_4$, $a_2 < a_3$, $a_5 < a_7$, $a_6 < a_7$, and the condition of a stable set of axioms is satisfied.

The importance of the concept of stable sets of axioms stems from the following theorem.

Theorem 2.5.3. Let Θ be any ep-signature, $\mathfrak{A}$ any stable set of elementary implications, and $f: A \longrightarrow B$ any injective homomorphism with $B \models \mathfrak{A}$. Then $A \models \mathfrak{A}$ holds, too.

Proof. By induction we will show that

$$A \models (v: G \longrightarrow t_1 \overset{e}{=} t_2)$$

holds for every elementary implication $(v: G \longrightarrow t_1 \overset{e}{=} t_2)$ of $\mathfrak{A}$. Thus, we assume that $(v: G \longrightarrow t_1 \overset{e}{=} t_2) \in \mathfrak{A}$ and that $A \models (u: H \longrightarrow r_1 \overset{e}{=} r_2)$ holds for every $(u: H \longrightarrow r_1 \overset{e}{=} r_2) \in \mathfrak{A}$ less than but not equal to the given elementary implication. It may be that no such implication exists. This is the case for minimal axioms. Since $\mathfrak{A}$ is stable it follows that

$$A \models (v: G \longrightarrow \mathrm{def}(t_1 \overset{e}{=} t_2))\ .$$

Now let $\underline{a} \in A_{(v:G)}$ and $s \in S$ with $t_1, t_2 \in T(\Sigma, v)_s$. Then $\underline{a} \in \mathrm{dom}\, t_1^A \cap \mathrm{dom}\, t_2^A$ because $A \models (v: G \longrightarrow \mathrm{def}(t_1 \overset{e}{=} t_2))$.
If we assume $\underline{a}\, t_1^A \neq \underline{a}\, t_2^A$, the injectivity of $f: A \longrightarrow B$ yields

$$\underline{a}\, t_1^A f_s \neq \underline{a}\, t_2^A f_s\ .$$

Since $f: A \longrightarrow B$ is not only an injective S-function but also a homomorphism, this inequality causes

$$\underline{a} f_v t_1^B = \underline{a}\, t_1^A f_s \neq \underline{a}\, t_2^A f_s = \underline{a} f_v t_2^B\ .$$

On the other hand $B \models (v: G \longrightarrow t_1 \overset{e}{=} t_2)$ holds because $B \models \mathfrak{A}$.
For $\underline{a} \in A_{(v:G)}$ this leads to

$$\underline{a} f_v \in B_{(v:G)} \subseteq B_{(v:\, t_1 \overset{e}{=} t_2)}\ .$$

But this contradicts

$$\underline{a} f_v t_1^B \neq \underline{a} f_v t_2^B\ ,$$

and hence

$$\underline{a}\, t_1^A = \underline{a}\, t_2^A$$

is true for every $\underline{a} \in A_{(v:G)}$. This means $A \models (v: G \longrightarrow t_1 \overset{e}{=} t_2)$, and the proof is finished. ∎

The following corollary shows how the concept of stable sets of axioms is related to the well-behaviour of homomorphisms between $(\Theta, \mathfrak{A})$-algebras. The expression $(\Theta, \mathfrak{A})$-algebra denotes Θ-algebras satisfying all axioms of $\mathfrak{A}$.

Corollary 2.5.4. For any ep-signature Θ and any stable set $\mathcal{A}$ of elementary implications the following conditions are equivalent:

(i) Θ is hierarchical;

(ii) Every bijective homomorphism $f: A \longrightarrow B$ between $(\Theta,\mathcal{A})$-algebras is an isomorphism;

(iii) Every injective homomorphism $f: A \longrightarrow B$ between $(\Theta,\mathcal{A})$-algebras reflects solutions of existence equations;

(iv) Every injective homomorphism $f: A \longrightarrow B$ between $(\Theta,\mathcal{A})$-algebras reflects solutions of domain conditions.

Proof. If we compare Theorem 2.4.2 and its proof, we see that we have only to extend the proof of (ii) $\Longrightarrow$ (i). Corollary 2.5.4 is proved if we can show that the bijective homomorphism $e: F^{*} \longrightarrow F$ as constructed in the proof on (2.4.2) is a homomorphism between $(\Theta,\mathcal{A})$-algebras. Obviously, the Θ-algebra F is a $(\Theta,\mathcal{A})$-algebra. Since $e: F^{*} \longrightarrow F$ is a bijective homomorphism and $\mathcal{A}$ is a stable set of axioms, $F^{*} \models \mathcal{A}$ holds due to Theorem 2.5.3. ∎

The results proved up to this point permit us to suggest hep-signatures with stable sets of elementary implications as a proper formalism to deal with quasi-varieties of partial algebras.

Definition 2.5.5. An ordered pair $T = (\Theta,\mathcal{A})$ consisting of a hep-signature Θ and a stable set $\mathcal{A}$ of elementary implications will be referred to as *hep-theory*.
The class of all those Θ-equoids which satisfies all axioms of $\mathcal{A}$ is called a *hep-quasi-variety* and will be denoted by $ALG(\Theta,\mathcal{A})$ or by $ALG(T)$. ∎

If we consider the set $\mathcal{A}_{CAT}$ which defines the class of all small categories, then this set of axioms is not only stable but it satisfies a further interesting property. For every elementary implication in $\mathcal{A}_{CAT}$ the domain condition of the conclusion is semantically equivalent to the premise, and this equivalence is a consequence of the set of axioms which are less than but not equal to the given implication. This kind of axioms corresponds to our understanding of equations for partial algebras, so we adopt the following convention.

Definition 2.5.6. A stable set of elementary implications $\mathcal{A}$ is called a set of *hep-equations*, if with respect to the required partial order $\leq$ on $\mathcal{A}$ the following condition is satisfied:

For every $(v\colon G \longrightarrow t_1 \stackrel{e}{=} t_2) \in \mathcal{A}$ the implication $(v\colon \mathrm{def}(t_1 \stackrel{e}{=} t_2) \longrightarrow G)$ is a consequence of

$$\{(u\colon H \longrightarrow r_1 \stackrel{e}{=} r_2) \in \mathcal{A} \mid (u\colon H \longrightarrow r_1 \stackrel{e}{=} r_2) \lneq (v\colon G \longrightarrow t_1 \stackrel{e}{=} t_2)\} .$$

The class of all equoids satisfying all axioms of a set of hep-equations will be called a h e p - v a r i e t y . ∎

As mentioned above, the class of all small categories constitutes a hep-variety.

It is not hard to see that hep-varieties and hep-quasi-varieties are closed with respect to subequoids and direct products. But we are not going to study closure properties of hep-varieties and hep-quasi-varieties in detail here or to prove Birkhoff-theorems. All that can be derived from corresponding theorems of the basic theory of partial algebras as developed in Part I by P. Burmeister.

CHAPTER 3

PARTIAL ALGEBRAS DEFINED BY GENERATORS AND RELATIONS

A motive for the development of a theory of partial algebras is the extension of the principle of structural induction from total algebras to partial ones. However, interesting free partial algebras are only those that are free with respect to some restricted class of partial algebras. Therefore we start this chapter with an investigation of sets of elementary implications.

In a similar way as for total algebras free partial algebras can be constructed from sets of terms by factorization. This construction requires a syntactic characterization of consequences of a set of axioms. In Section 3.1 we give a complete set of rules of derivation for elementary implications and prove a completeness theorem.

The main notion of Section 3.2 is the notion of a partial algebra $F(\mathcal{A},G,v)$ freely generated by a set of existence equations $(v:G)$ with respect to a set $\mathcal{A}$ of elementary implications serving as axioms. This notion extends the well-known concept of the theory of groups defining a group by generators and relations. In the case of the theory of groups the set of axioms $\mathcal{A}$ defines the class of groups, the system of variables $v: X \longrightarrow S$ represents the set of generators, and $(v:G)$ represents the set of defining relations. This concept comprises the concept of an $\mathcal{A}$-algebra freely generated by an S-set, if we take for $(v:G)$ the empty set of existence equations $(v:\emptyset)$.

By means of the concept of a partial algebra freely generated by a set of existence equations we can give the construction of partial quotient algebras, of free sums of partial algebras, sometimes called coproducts, and of other constructions of colimits.These constructions can be carried out by constructions with defining sets of existence equations.These constructions will be done in Section 3.3.

In the next section we use the construction of partial quotient algebras for an extension of the homomorphism theorem. The homomorphism theorem of total algebras says that a homomorphic image can be reconstructed from the domain and the kernel of a homomorphism. We show that in general the homomorphic image of a homomorphism between equoids cannot be reconstructed in a finite way from the domain and the kernel of the homomorphism. But we will see that the homomorphic image is the directed limit of the chain of iterated quotient equoids starting from the domain of the homomorphism. The case of total algebras is then characterized by the fact that the chain of iterated quotient equoids is of length one.

In the last section of this chapter we deal with the construction of relatively free equoids. This construction becomes the algebraic basis of the last chapter.

3.1 Derivability of elementary implications

A first version of a set of rules of derivation for elementary implications was given in Kaphengst,Reichel (1971). A completeness theorem for first order languages on homogeneous partial algebras was given by Thiele (1966), however this was not noticed by algebraists. It is worth to mention that rules of derivation of homogeneous algebras cannot be extended directly to many-sorted algebras. This problem is thoroughly discussed in Goguen,Meseguer (1981).

We will give a complete set of rules of derivation that allows us to describe syntactically that an existence equation $(v\colon t_1 \overset{e}{=} t_2)$ is a consequence of a set of existence equations $(v\colon G)$ with respect to a given set $\mathfrak{A}$ of axioms. The system of rules of derivation is a very condensed one since there is only one rule that can be applied iteratively. This system was published first in Kaphengst (1981).

For the sake of simplicity we are working with partial Σ-algebras for any signature $\Sigma = (S, \alpha\colon \Omega \longrightarrow S^{*} \times S)$. The concept of hep-signatures is not of importance for this section.

<u>Definition 3.1.1</u>. Let $\Sigma = (S, \alpha\colon \Omega \longrightarrow S^{*} \times S)$ be any signature. For every $s \in S$, $v_s\colon \{x,y,z\} \longrightarrow S$ may be the S-sorted system of variables with $x v_s = y v_s = z v_s = s$. Then the following elementary implications are called t a u t o l o g i c a l :

$REF_s\colon\ (v_s\colon \emptyset \longrightarrow x \overset{e}{=} x),\quad s \in S$

$SYM_s\colon\ (v_s\colon x \overset{e}{=} y \longrightarrow y \overset{e}{=} x),\quad s \in S$

$TRA_s\colon\ (v_s\colon x \overset{e}{=} y,\ y \overset{e}{=} z \longrightarrow x \overset{e}{=} z),\quad s \in S$

For every $\sigma \in \Omega$ with $\sigma\colon w \longrightarrow s$ and $w\colon [n] \longrightarrow S$ and for every $j \in [n]$ we define the S-sorted system of variables

$$v_{\sigma,j} : \{x_1,\dots,x_n,y,z\} \longrightarrow S$$

by $x_i v_{\sigma,j} = iw$ for every $i \in [n]$, and $y v_{\sigma,j} = jw$, $z v_{\sigma,j} = s$.
With this S_sorted system of variables we define further tautological elementary implications:

$$SUB_{\sigma,j} : \; (v_{\sigma,j} : \; x_j \overset{e}{=} y \,,\, z \overset{e}{=} (x_1,\dots,x_n)\sigma \longrightarrow z \overset{e}{=} (x_1,\dots,x_{j-1},y,x_{j+1},\dots,x_n)\sigma \;) .$$

Finally we call

$$TAUT(\Sigma) = \{ REF_s \mid s \in S \} \cup \{ SYM_s \mid s \in S \} \cup \{ TRA_s \mid s \in S \} \cup \cup \{ SUB_{\sigma,j} \mid \sigma \in \Omega ,\, \sigma : w \longrightarrow s \,,\, w : [n] \longrightarrow S \,,\, j \in [n] \}$$

the set of tautological elementary implications of Σ. ∎

These elementary implications are called tautological since

$$\emptyset \models (v : G \longrightarrow t_1 \overset{e}{=} t_2)$$

holds for every

$$(v : G \longrightarrow t_1 \overset{e}{=} t_2) \in TAUT(\Sigma) .$$

<u>Definition 3.1.2</u>. An existence equation $(v : t_1 \overset{e}{=} t_2)$ with $v : \{x_1,\dots,x_n\} \longrightarrow S$ is said to be $\mathcal{A}$- d e r i v a b l e from a finite set of existence equations $(v{:}G)$, denoted by

$$\mathcal{A} \vdash (v : G \longrightarrow t_1 \overset{e}{=} t_2) ,$$

where $\mathcal{A}$ is any set of elementary implications, if it is derivable using only the following rules of derivation DI, DII, DIII:

DI: $\mathcal{A} \vdash (v : G \longrightarrow t_1 \overset{e}{=} t_2)$ for every $(v : t_1 \overset{e}{=} t_2) \in (v{:}G)$;

DII: $\mathcal{A} \vdash (v : G \longrightarrow x \overset{e}{=} x)$ for every $x \in X$ with $v : X \longrightarrow S$;

DIII: If

(a) $(w : H \longrightarrow t_1 \overset{e}{=} t_2) \in TAUT(\Sigma) \cup \mathcal{A}$ with $w : Y \longrightarrow S$,

(b) $\underline{t} \in T(\Sigma, v)_w$ such that for every $y \in Y$ an existence equation $(v : r_1 \overset{e}{=} r_2)$ exists so that $\mathcal{A} \vdash (v : G \longrightarrow r_1 \overset{e}{=} r_2)$ and $y\underline{t}$ is a subterm of r_1 or r_2, and if

(c) $\mathcal{A} \vdash (v : G \longrightarrow p_1\underline{\tilde{t}} \overset{e}{=} p_2\underline{\tilde{t}})$ holds for every $(w : p_1 \overset{e}{=} p_2) \in (w{:}H)$

then follows

$$\mathcal{A} \vdash (v : G \longrightarrow t_1\underline{t} \overset{e}{=} t_2\underline{t}) .$$

$IMP(\mathfrak{A})$ denotes the set of all elementary implications $(v: G \longrightarrow t_1 \overset{e}{=} t_2)$ with $\mathfrak{A} \vdash (v: G \longrightarrow t_1 \overset{e}{=} t_2)$. ■

To improve the understanding of this definition we give another description of the set $IMP(\mathfrak{A})$. The rule of derivation DIII is applicable only if there are any $\mathfrak{A}$-derivable elementary implications satisfying (a), (b), and (c). On the other hand, by means of DI and DII one can derive elementary implications without any prerequisites. Therefore we denote by

$$IMP_o(\mathfrak{A})$$

the set of all those elementary implications that are $\mathfrak{A}$-derivable only by means of DI and DII. This means

$$IMP_o(\mathfrak{A}) = \left\{(v: G \longrightarrow t_1 \overset{e}{=} t_2) \mid (v: t_1 \overset{e}{=} t_2) \in (v:G)\right\} \cup$$
$$\cup \left\{(v: G \longrightarrow x \overset{e}{=} x) \mid v: X \longrightarrow S,\ x \in X\right\}.$$

Since DIII is the only rule of derivation that can be applied iteratively, we set

$IMP_{i+1}(\mathfrak{A}) = IMP_i(\mathfrak{A})$ joined with the set of all those elementary implications $(v: G \longrightarrow t_1 \underline{\tilde{t}} \overset{e}{=} t_2 \underline{\tilde{t}})$ that are $\mathfrak{A}$-derivable by DIII in such a way that in (b) for every $y \in Y$ an existence equation $(v: r_1 \overset{e}{=} r_2)$ exists with $(v: G \longrightarrow r_1 \overset{e}{=} r_2) \in IMP_i(\mathfrak{A})$ and $y\underline{t}$ is a subterm of r_1 or r_2, and that with respect to (c) $(v: G \longrightarrow p_1 \underline{\tilde{t}} \overset{e}{=} p_2 \underline{\tilde{t}}) \in IMP_i(\mathfrak{A})$ holds for every $(w: p_1 \overset{e}{=} p_2) \in (w:H)$.

Thus, $IMP_i(\mathfrak{A})$ is the set of all elementary implications which are $\mathfrak{A}$-derivable by i-fold application of the rule DIII. This leads to

$$IMP(\mathfrak{A}) = \bigcup_{i=0}^{\infty} IMP_i(\mathfrak{A}).$$

Naturally, everybody expects that every tautological elementary implication and every elementary implication of $\mathfrak{A}$ is $\mathfrak{A}$-derivable. But this follows not immediately from Definition 3.1.2. For this reason we prove the following corollary.

Corollary 3.1.3. For every set of elementary implications

$$\mathfrak{A} \cup TAUT(\Sigma) \subseteq IMP_1(\mathfrak{A}) \subseteq IMP(\mathfrak{A})$$

holds.

__Proof__. According to the preceding consideration we have to show

$$\mathcal{A} \cup \mathrm{TAUT}(\Sigma) \subseteq \mathrm{IMP}_1(\mathcal{A}).$$

Therefore, let $(w\colon H \longrightarrow t_1 \overset{\circ}{=} t_2)$ be in $\mathrm{TAUT}(\Sigma) \cup \mathcal{A}$ with $w\colon Y \longrightarrow S$, and let $\underline{t} \in T(\Sigma, w)_w$ with $y\underline{t} = y$ for every $y \in Y$. Because of DII, $\mathcal{A} \vdash (w\colon H \longrightarrow y \overset{\circ}{=} y)$ holds for every $y \in Y$ and even $(w\colon H \longrightarrow y \overset{\circ}{=} y) \in \mathrm{IMP}_0(\mathcal{A})$ for every $y \in Y$. Evidently, $y\underline{t} = y$ is a subterm of y so that $\underline{t} \in T(\Sigma, w)_w$ satisfies condition (b) of DIII. The rule of derivation DI implies

$$\mathcal{A} \vdash (w\colon H \longrightarrow p_1 \overset{\circ}{=} p_2) \text{ for every } (w\colon p_1 \overset{\circ}{=} p_2) \in (w\colon H),$$

and so

$$(w\colon H \longrightarrow p_1 \overset{\circ}{=} p_2) \in \mathrm{IMP}_0(\mathcal{A}).$$

Since $\tilde{\underline{t}}$ is the identity of $T(\Sigma, w)$ we obtain

$$\mathcal{A} \vdash (w\colon H \longrightarrow t_1 \overset{\circ}{=} t_2) \text{ and } (w\colon H \longrightarrow t_1 \overset{\circ}{=} t_2) \in \mathrm{IMP}_1(\mathcal{A})$$

by rule DIII. ∎

Another useful property of $\mathcal{A}$-derivability is given by the following corollary.

__Corollary 3.1.4__. If $\mathcal{A} \vdash (v\colon G \longrightarrow t_1 \overset{\circ}{=} t_2)$, and if $r \in T(\Sigma, v)_s$, $s \in S$, is a subterm of t_1 or t_2, then $\mathcal{A} \vdash (v\colon G \longrightarrow r \overset{\circ}{=} r)$.

__Proof__. If we take $(v_s\colon \emptyset \longrightarrow x \overset{\circ}{=} x) \in \mathrm{TAUT}(\Sigma)$ and $\underline{t}\colon \{x, y, z\} \longrightarrow T(\Sigma, v)_s$ with $x\underline{t} = y\underline{t} = z\underline{t} = r$, then conditions (b) and (c) are satisfied with respect to $(v\colon G)$, so we obtain

$$\mathcal{A} \vdash (v\colon G \longrightarrow x\tilde{\underline{t}} \overset{\circ}{=} x\tilde{\underline{t}}) \text{ or } \mathcal{A} \vdash (v\colon G \longrightarrow r \overset{\circ}{=} r)$$

by DIII. ∎

As next we formulate the g e n e r a l i z e d s u b s t i t u t i o n r u l e which will be referred to by GSR.

__Corollary 3.1.5__. If $v\colon X \longrightarrow S$, $w\colon Y \longrightarrow S$ are finite S-sorted systems of variables, if $\underline{t}_1, \underline{t}_2 \in T(\Sigma, v)_w$ are assignments, and if $t \in T(\Sigma, w)_s$, $s \in S$, then from

$$\mathcal{A} \vdash (v\colon G \longrightarrow y\tilde{\underline{t}}_1 \overset{\circ}{=} y\tilde{\underline{t}}_2) \text{ for every } y \in Y, \text{ and}$$

$$\mathcal{A} \vdash (v\colon G \longrightarrow t\tilde{\underline{t}}_1 \overset{\circ}{=} t\tilde{\underline{t}}_1)$$

follows

$$\mathcal{A} \vdash (v\colon G \longrightarrow t\tilde{\underline{t}}_1 \overset{\circ}{=} t\tilde{\underline{t}}_2).$$

Proof. GSR will be proved by induction on the structure of $t \in T(\Sigma,w)$.

If $t = y \in T(\Sigma,w)_{yw}$ or $t = \sigma^{T(\Sigma,w)}$ with $\sigma : \lambda \longrightarrow s$, i.e. if $\sigma \in \Omega$ is a constant operator, then the conclusion of GSR coincides with one of the premises.

Corresponding to the inductive proof we assume that $t = \underline{p}\sigma^{T(\Sigma,w)}$ with $\sigma : u \longrightarrow s$, $u:[n] \longrightarrow S$, $\underline{p} \in T(\Sigma,w)_u$, and that GSR is applicable to $i\underline{p}$ for every $i \in [n]$.
Since

$$(\underline{p}\sigma^{T(\Sigma,w)})\tilde{\underline{t}}_1 = \underline{p}\tilde{\underline{t}}_1\sigma^{T(\Sigma,v)}$$

the premises of GSR yield

$$\mathfrak{A} \vdash (v: G \longrightarrow \underline{p}\tilde{\underline{t}}_1\sigma^{T(\Sigma,v)} \overset{e}{=} \underline{p}\tilde{\underline{t}}_1\sigma^{T(\Sigma,v)}) \text{ and}$$

$$\mathfrak{A} \vdash (v: G \longrightarrow y\tilde{\underline{t}}_1 \overset{e}{=} y\tilde{\underline{t}}_2) \text{ for every } y \in Y.$$

Now we take the tautological elementary implication $SUB_{\sigma,1}$, defined in (3.1.1), and

$$\underline{r} \in T(\Sigma,v)_{v_{\sigma,1}} \quad \text{with } x_j\underline{r} = (j\underline{p})\tilde{\underline{t}}_1 \text{ for } j \in [n],$$

with

$$z\underline{r} = \underline{p}\tilde{\underline{t}}_1\sigma^{T(\Sigma,v)}, \quad y\underline{r} = (1\underline{p})\tilde{\underline{t}}_2,$$

and we check conditions (b) and (c) of DIII with respect to $SUB_{\sigma,1}$,

$$\underline{r} \in T(\Sigma,v)_{v_{\sigma,1}}$$

and with respect to $(v:G)$.
Since GSR is applicable to $1\underline{p}$,

$$\mathfrak{A} \vdash (v: G \longrightarrow (1\underline{p})\tilde{\underline{t}}_1 \overset{e}{=} (1\underline{p})\tilde{\underline{t}}_2)$$

follows. This together with

$$\mathfrak{A} \vdash (v: G \longrightarrow \underline{p}\tilde{\underline{t}}_1\sigma^{T(\Sigma,v)} \overset{e}{=} \underline{p}\tilde{\underline{t}}_1\sigma^{T(\Sigma,v)})$$

guarantees that for every $x \in \{y,z,x_1,\dots,x_n\}$ there is an existence equation $(v: q_1 \overset{e}{=} q_2)$ with

$$\mathfrak{A} \vdash (v: G \longrightarrow q_1 \overset{e}{=} q_2)$$

so that $x\underline{r}$ is a subterm of q_1 or q_2.
Condition (c) of DIII is satisfied because

$$\mathfrak{A} \vdash (v: G \longrightarrow (1\underline{p})\tilde{\underline{t}}_1 \overset{e}{=} (1\underline{p})\tilde{\underline{t}}_2),$$

$$\mathfrak{A} \vdash (v: G \longrightarrow \underline{p}\tilde{\underline{t}}_1\sigma^{T(\Sigma,v)} \overset{e}{=} \underline{p}\tilde{\underline{t}}_1\sigma^{T(\Sigma,v)}), \text{ and}$$

$$((x_1,\dots,x_n)\sigma^{T(\Sigma,v_{\sigma,1})})\tilde{\underline{r}} = (x_1\underline{r},\dots,x_n\underline{r})\sigma^{T(\Sigma,v)} \;\blacksquare$$

$= \underline{p}\,\tilde{\underline{t}}_1 \sigma^{T(\Sigma, v)}$.

By means of DIII we get

$$\mathcal{A} \vdash (v\colon G \longrightarrow z\,\underline{r} \overset{e}{=} ((y, x_2, \dots, x_n)\,\sigma^{T(\Sigma, v_{\sigma,1})})\,\tilde{\underline{r}})$$

and so

$$\mathcal{A} \vdash (v\colon G \longrightarrow \underline{p}\,\tilde{\underline{t}}_1 \sigma^{T(\Sigma, v)} \overset{e}{=} ((1\underline{p})\,\tilde{\underline{t}}_1, \dots, (n\underline{p})\,\tilde{\underline{t}}_1)\,\sigma^{T(\Sigma, v)}) .$$

By an (n-1)-fold iteration of the preceding step one can finally prove

$$\mathcal{A} \vdash (v\colon G \longrightarrow t\,\tilde{\underline{t}}_1 \overset{e}{=} t\,\tilde{\underline{t}}_2) . \qquad \blacksquare$$

Finally we want to extend the rule of derivation DIII in such a way that in the premise (a) not only axioms or tautological elementary implications are allowed but arbitrary $\mathcal{A}$-derivable elementary implications.

Corollary 3.1.6. $IMP(\mathcal{A})$ is closed with respect to the following extended rule of derivation EDIII:

If

(a) $\mathcal{A} \vdash (w\colon H \longrightarrow t_1 \overset{e}{=} t_2)$, $w\colon Y \longrightarrow S$,

(b) $\underline{t} \in T(\Sigma, v)_w$ such that for every $y \in Y$ there is an existence equation $(v\colon r_1 \overset{e}{=} r_2)$ with $\mathcal{A} \vdash (v\colon G \longrightarrow r_1 \overset{e}{=} r_2)$ so that $y\underline{t}$ is a subterm of r_1 or r_2 , and if

(c) $\mathcal{A} \vdash (v\colon G \longrightarrow p_1 \tilde{\underline{t}} \overset{e}{=} p_2 \tilde{\underline{t}})$ for every $(w\colon p_1 \overset{e}{=} p_2) \in (w\colon H)$,

then

$$\mathcal{A} \vdash (v\colon G \longrightarrow t_1 \tilde{\underline{t}} \overset{e}{=} t_2 \tilde{\underline{t}}) .$$

Proof. Corollary 3.1.6 will be proved by use of the inductive representation

$$IMP(\mathcal{A}) = \bigcup_{i=0}^{\infty} IMP_i(\mathcal{A}) .$$

First we show that EDIII is correct if in (a) an elementary implication of $IMP_o(\mathcal{A})$ is used.

Let

$$(w\colon H \longrightarrow t_1 \overset{e}{=} t_2) \in IMP_o(\mathcal{A}) \text{ with } w\colon Y \longrightarrow S , \underline{t} \in T(\Sigma, v)_w ,$$

and let $(v\colon G)$ be a set of existence equations such that (b) and (c) are satisfied. On account of the definition of $IMP_o(\mathcal{A})$ two cases are to be considerd. One case is given by $(w\colon t_1 \overset{e}{=} t_2) \in (w\colon H)$ and the other one by $t_1 = t_2 = y_o \in Y$.

If $(w\colon t_1 \overset{e}{=} t_2) \in (w\colon H)$, then condition (c) implies

$$\mathcal{A} \vdash (v\colon G \longrightarrow p_1 \tilde{\underline{t}} \overset{e}{=} p_2 \tilde{\underline{t}}) \text{ for all } (w\colon p_1 \overset{e}{=} p_2) \in (w\colon H) ,$$

and so

$$\mathfrak{A} \vdash (v: G \longrightarrow t_1\underline{\tilde{t}} \overset{e}{=} t_2\underline{\tilde{t}}).$$

Now let $t_1 = t_2 = y_0 \in Y$ and $s = y_0 w$. Then condition (b) together with Corollary 3.1.4 yields

$$\mathfrak{A} \vdash (v: G \longrightarrow y\underline{t} \overset{e}{=} y\underline{t}) \quad \text{for every} \quad y \in Y.$$

This leads to

$$\mathfrak{A} \vdash (v: G \longrightarrow t_1\underline{\tilde{t}} \overset{e}{=} t_2\underline{\tilde{t}})$$

because $t_1\underline{\tilde{t}} = y_0\underline{t} = y_0\underline{t} = t_2\underline{\tilde{t}}$.

Now, according to the induction, we assume that EDIII produces $\mathfrak{A}$-derivable elementary implications, provided in (a) an elementary implication of $IMP_i(\mathfrak{A})$, $i \geq 0$, is used. We have to show by means of this assumption that EDIII also produces $\mathfrak{A}$-derivable elementary implications if in (a) an elementary implication of $IMP_{i+1}(\mathfrak{A})$ is used.

Let be given

$$(w: H \longrightarrow t_1 \overset{e}{=} t_2) \in IMP_{i+1}(\mathfrak{A}), \; w: Y \longrightarrow S, \; \underline{t} \in T(\Sigma,v)_w,$$

$v: X \longrightarrow S$, and $(v:G)$ such that (b) and (c) of EDIII are satisfied. Because of

$$(w: H \longrightarrow t_1 \overset{e}{=} t_2) \in IMP_{i+1}(\mathfrak{A})$$

there is an elementary implication

$$(u: H^* \longrightarrow q_1 \overset{e}{=} q_2) \in TAUT(\Sigma) \cup \mathfrak{A}, \quad u: Z \longrightarrow S,$$

and there is an assignment $\underline{r} \in T(\Sigma,w)_u$ such that (b) and (c) of DIII are satisfied with respect to $(w:H)$, and furthermore $t_1 = q_1\underline{\tilde{r}}$, $t_2 = q_2\underline{\tilde{r}}$.

Thus, we have an implication

$$(u: H^* \longrightarrow q_1 \overset{e}{=} q_2) \in TAUT(\Sigma) \cup \mathfrak{A}$$

and an assignment $\underline{r}\underline{\tilde{t}} \in T(\Sigma,v)_u$ which satisfy (b) and (c) of DIII with respect to $(v:G)$, what will be proved in the following.

First we consider condition (b):
For every $z \in Z$ there is an existence equation $(w: p_1 \overset{e}{=} p_2)$ with

$$(w: H \longrightarrow p_1 \overset{e}{=} p_2) \in IMP_i(\mathfrak{A})$$

so that $z\underline{r}$ is a subterm of p_1 or p_2.
According to our assumption we can apply EDIII to $(w: H \longrightarrow p_1 \overset{e}{=} p_2)$.
So for every $z \in Z$ there is an existence equation $(v: p_1\underline{\tilde{t}} = p_2\underline{\tilde{t}})$ with

$$\mathfrak{A} \vdash (v: G \longrightarrow p_1\underline{\tilde{t}} = p_2\underline{\tilde{t}})$$

so that $(z\,\underline{r})\underline{\tilde{t}}$ is a subterm of $p_1\underline{\tilde{t}}$ or $p_2\underline{\tilde{t}}$, because $z\,\underline{r}$ is a subterm of p_1 or p_2, and

$$\underline{\tilde{t}} : T(\Sigma, w) \longrightarrow T(\Sigma, v)$$

preserves the subterm relation. Consequently, (b) is satisfied.

Next we consider condition (c), i.e. we have to prove that

$$\mathfrak{A} \vdash (v: G \longrightarrow p_1(\widetilde{\underline{r}\,\underline{\tilde{t}}}) \overset{e}{=} p_2(\widetilde{\underline{r}\,\underline{\tilde{t}}}))$$

holds for every $(u: p_1 \overset{e}{=} p_2) \in (u: H^*)$.

The required properties of $\underline{r} \in T(\Sigma, w)_u$ and

$$(u: H^* \longrightarrow q_1 \overset{e}{=} q_2) \in TAUT(\Sigma) \cup \mathfrak{A}$$

yield

$$(w: H \longrightarrow p_1\underline{\tilde{r}} \overset{e}{=} p_2\underline{\tilde{r}}) \in IMP_1(\mathfrak{A}) \quad \text{for every} \quad (u: p_1 \overset{e}{=} p_2) \in (u: H^*).$$

Hence we can apply EDIII to each of these elementary implications and we obtain

$$\mathfrak{A} \vdash (v: G \longrightarrow (p_1\underline{\tilde{r}})\underline{\tilde{t}} \overset{e}{=} (p_2\underline{\tilde{r}})\underline{\tilde{t}})$$

for every $(u: p_1 \overset{e}{=} p_2) \in (u: H^*)$.

Condition (c) is also satisfied because

$$p_1(\widetilde{\underline{r}\,\underline{\tilde{t}}}) = (p_1\underline{\tilde{r}})\underline{\tilde{t}} \quad \text{and} \quad p_2(\widetilde{\underline{r}\,\underline{\tilde{t}}}) = (p_2\underline{\tilde{r}})\underline{\tilde{t}}.$$

Now we can apply DIII to $(u: H^* \longrightarrow q_1 \overset{e}{=} q_2) \in TAUT(\Sigma) \cup \mathfrak{A}$, $\underline{r}\,\underline{\tilde{t}} \in T(\Sigma, v)_u$ and $(v: G)$, and we obtain

$$\mathfrak{A} \vdash (v: G \longrightarrow q_1(\widetilde{\underline{r}\,\underline{\tilde{t}}}) \overset{e}{=} q_2(\widetilde{\underline{r}\,\underline{\tilde{t}}})).$$

Because $q_1(\widetilde{\underline{r}\,\underline{\tilde{t}}}) = (q_1\underline{\tilde{r}})\underline{\tilde{t}} = t_1\underline{\tilde{t}}$ and $q_2(\widetilde{\underline{r}\,\underline{\tilde{t}}}) = t_2\underline{\tilde{t}}$ we have proved

$$(v: G \longrightarrow t_1\underline{\tilde{t}} \overset{e}{=} t_2\underline{\tilde{t}}) \in IMP(\mathfrak{A}).$$ ■

The next definition extends the concept of $\mathfrak{A}$-derivability from elementary implications to arbitrary implications. As preparation we consider more thoroughly terms on infinite S-sorted systems of variables. Since every operator $\sigma \in \Omega$ has a finite arity $\sigma : w \longrightarrow s$ with $w: [n] \longrightarrow S$, for every term $t \in T(\Sigma, v)$, where $v: X \longrightarrow S$ is an arbitrary S-sorted system of variables, there exists a finite subsystem $w: Y \longrightarrow S$ of $v: X \longrightarrow S$ with

$$t \in T(\Sigma, w) \subseteq T(\Sigma, v).$$

We say that $w: Y \longrightarrow S$ is a finite subsystem of $v: X \longrightarrow S$, if Y is a finite subset of X and $yw = yv$ for every $y \in Y$.

If $t \in T(\Sigma, v)$ and $w: Y \longrightarrow S$ are related in this way, then we say that w c o v e r s t, denoted by

$$w \succ t .$$

We say that a finite S-sorted system of variables $w: Y \longrightarrow S$ covers the existence equation $(v: t_1 \overset{e}{=} t_2)$ or the set of existence equations $(v:G)$, if w covers as well t_1 as t_2 or if w covers every existence equation of $(v:G)$, respectively. These facts will be denoted by

$$w \succ (v: t_1 \overset{e}{=} t_2) \quad \text{or} \quad w \succ (v:G),$$

respectively.

If $w \succ (v: t_1 \overset{e}{=} t_2)$, or $w \succ (v:G)$, with $v: X \longrightarrow S$, $w: Y \longrightarrow S$, then we denote by

$$(v{\downarrow}w: t_1 \overset{e}{=} t_2) \quad \text{or} \quad (v{\downarrow}w: G)$$

the corresponding existence equation or set of existence equations on the domain $w: Y \longrightarrow S$, which results by forgetting all those variables of X that are not contained in Y. If we use the notation $(v{\downarrow}w: t_1 \overset{e}{=} t_2)$ or $(v{\downarrow}w: G)$, then we state simultaneously

$$w \succ (v: t_1 \overset{e}{=} t_2) \quad \text{or} \quad w \succ (v: G),$$

respectively.

Definition 3.1.7. Let $\mathcal{A}$ be any set of elementary implications and $v: X \longrightarrow S$ any S-sorted system of variables. A set of existence equations $(v: H)$ is said to be $\mathcal{A}$ - d e r i v a b l e from a set of existence equations $(v: G)$, denoted by

$$\mathcal{A} \vdash (v: G \longrightarrow H),$$

if for every $(v: t_1 \overset{e}{=} t_2) \in (v: H)$ there exists a finite S-sorted subsystem $w: Y \longrightarrow S$ of $v: X \longrightarrow S$ and a finite subset $(v: G^*)$ of $(v: G)$ such that

$$w \succ (v: G^*), \quad w \succ (v: t_1 \overset{e}{=} t_2) \quad \text{and}$$

$$\mathcal{A} \vdash (v{\downarrow}w: G^* \longrightarrow t_1 \overset{e}{=} t_2) .$$

■

A useful property of $\mathcal{A}$-derivability of implications describes the following corollary.

Corollary 3.1.8. If $(w: F)$, $(w: H)$ are sets of existence equations with $w: Y \longrightarrow S$ and $H \neq \emptyset$, if $(v: G)$ with $v: X \longrightarrow S$ is a further set of existence equations, and if $t \in T(\Sigma, v)_w$ and $\mathcal{A}$ any set of elementary implications, then

$$\mathfrak{A} \vdash (v\colon G \longrightarrow H\widetilde{\underline{t}}) \quad \text{and} \quad \mathfrak{A} \vdash (w\colon H \longrightarrow F)$$

imply

$$\mathfrak{A} \vdash (v\colon G \longrightarrow F\widetilde{\underline{t}}) .$$

Proof. Let $(w\colon t_1 \overset{e}{=} t_2)$ be any existence equation of $(w\colon F)$, then $(v\colon t_1\widetilde{\underline{t}} \overset{e}{=} t_2\widetilde{\underline{t}})$ is an arbitrary existence equation of $(v\colon F\widetilde{\underline{t}})$.

Because $\mathfrak{A} \vdash (w\colon H \longrightarrow F)$ there exists a finite S-sorted subsystem $u\colon Z \longrightarrow S$ of $w\colon Y \longrightarrow S$ and a finite subset $H^* \subseteq H$ with $u \succ\!\!- (w\colon H^*)$, $u \succ\!\!- (w\colon t_1 \overset{e}{=} t_2)$ and $\mathfrak{A} \vdash (w{\downarrow}u\colon H^* \longrightarrow t_1 \overset{e}{=} t_2)$.

By means of EDIII this implies

$$\mathfrak{A} \vdash (v\colon H^*\widetilde{\underline{t}} \longrightarrow t_1\widetilde{\underline{t}} \overset{e}{=} t_2\widetilde{\underline{t}}) .$$

Because of $\mathfrak{A} \vdash (v\colon G \longrightarrow H\widetilde{\underline{t}})$ and of the finiteness of $(w\colon H^*)$ there exists a finite S-sorted subsystem $v^*\colon X^* \longrightarrow S$ of $v\colon X \longrightarrow S$ and a finite subset $G^* \subseteq G$ such that

$$v^* \succ\!\!- (v\colon G^*) , \quad v^* \succ\!\!- (v\colon H^*\widetilde{\underline{t}}), \quad v^* \succ\!\!- (v\colon t_1\widetilde{\underline{t}} \overset{e}{=} t_2\widetilde{\underline{t}}) \quad \text{and}$$

$$\mathfrak{A} \vdash (v{\downarrow}v^*\colon G^* \longrightarrow H^*\widetilde{\underline{t}}) .$$

From

$$\mathfrak{A} \vdash (v{\downarrow}v^*\colon G^* \longrightarrow H^*\widetilde{\underline{t}}) \text{ and } \mathfrak{A} \vdash (v\colon H^*\widetilde{\underline{t}} \longrightarrow t_1\widetilde{\underline{t}} \overset{e}{=} t_2\widetilde{\underline{t}}) ,$$

$$\mathfrak{A} \vdash (v{\downarrow}v^*\colon G^* \longrightarrow t_1\widetilde{\underline{t}} = t_2\widetilde{\underline{t}})$$

follows by means of EDIII.

Consequently, for every $(v\colon t_1\widetilde{\underline{t}} \overset{e}{=} t_2\widetilde{\underline{t}}) \in (v\colon F\widetilde{\underline{t}})$ there exists a finite S-sorted subsystem $v^*\colon X^* \longrightarrow S$ of $v\colon X \longrightarrow S$ and a finite $G^* \subseteq G$ with

$$v^* \succ\!\!- (v\colon G^*) , \quad v^* \succ\!\!- (v\colon t_1\widetilde{\underline{t}} \overset{e}{=} t_2\widetilde{\underline{t}}) \quad \text{and}$$

$$\mathfrak{A} \vdash (v\colon G^* \longrightarrow t_1\widetilde{\underline{t}} \overset{e}{=} t_2\widetilde{\underline{t}}) ,$$

which implies

$$\mathfrak{A} \vdash (v\colon G \longrightarrow F\widetilde{\underline{t}}) .$$ ∎

The proof of correctness or soundness concludes this section. The completeness of $\mathfrak{A}$-derivability will be proved in the following section.

Theorem 3.1.9. Let $\mathfrak{A}$ be any set of elementary implications and $(v\colon G \longrightarrow H)$ with $v\colon X \longrightarrow S$ any implication. For every $\mathfrak{A}$-algebra A the statement

$$\mathfrak{A} \vdash (v\colon G \longrightarrow H) \quad \text{implies} \quad A_{(v:G)} \subseteq A_{(v:H)} .$$

Proof. The proof is based again on the inductive representation

$$\mathrm{IMP}(\mathcal{A}) = \bigcup_{i=0}^{\infty} \mathrm{IMP}_i(\mathcal{A}).$$

First we show that $\mathcal{A}$-derivability is sound for elementary implications. Let

$$(w\colon G \longrightarrow t_1 \overset{e}{=} t_2) \in \mathrm{IMP}_0(\mathcal{A}),$$

then $w\colon Y \longrightarrow S$ is a finite S-sorted system of variables. Therefore either $(w\colon t_1 \overset{e}{=} t_2) \in (w\colon G)$ or $t_1 = t_2 = y \in Y$.
Because

$$A_{(w:G)} = \bigcap \{ A_{(w:t_1 \overset{e}{=} t_2)} \mid (w\colon t_1 \overset{e}{=} t_2) \in (w\colon G) \},$$

$$A_{(w:G)} \subseteq A_{(w:t_1 \overset{e}{=} t_2)}$$

holds for every $(w\colon t_1 \overset{e}{=} t_2) \in (w\colon G)$.
Since $(w\colon G)$ is a set of existence equations on $w\colon Y \longrightarrow S$,

$$A_{(w:G)} \subseteq A_w = A_{(w:y \overset{e}{=} y)}$$

holds for every $y \in Y$.

Now we assume that (3.1.9) is true for every elementary implication of $\mathrm{IMP}_i(\mathcal{A})$, $i \geq 0$.
If

$$(w\colon G \longrightarrow t_1 \overset{e}{=} t_2) \in \mathrm{IMP}_{i+1}(\mathcal{A}),$$

then there are

$$(u\colon H \longrightarrow r_1 \overset{e}{=} r_2) \in \mathrm{TAUT}(\Sigma) \cup \mathcal{A}, \quad u\colon Z \longrightarrow S, \quad \underline{t} \in T(\Sigma, w)_u$$

such that conditions (b) and (c) of DIII are satisfied with respect to $(w\colon G)$, and $r_1 \tilde{\underline{t}} = t_1$, $r_2 \tilde{\underline{t}} = t_2$.
Consequently, for every $z \in Z$ there is an existence equation $(w\colon q_1 \overset{e}{=} q_2)$ with

$$(w\colon G \longrightarrow q_1 \overset{e}{=} q_2) \in \mathrm{IMP}_i(\mathcal{A})$$

and $z\underline{t}$ is a subterm of q_1 or q_2. Additionally

$$(w\colon G \longrightarrow p_1 \tilde{\underline{t}} \overset{e}{=} p_2 \tilde{\underline{t}}) \in \mathrm{IMP}_i(\mathcal{A})$$

holds for every $(u\colon p_1 \overset{e}{=} p_2) \in (u\colon H)$.
According to the notion of an $\mathcal{A}$-algebra

$$A_{(u:H)} \subseteq A_{(u:r_1 \overset{e}{=} r_2)}$$

holds for every $(u\colon H \longrightarrow r_1 \overset{e}{=} r_2) \in \mathcal{A}$, and without any problems one can show that

$$A_{(u:H)} \subseteq A_{(u:r_1 \overset{e}{=} r_2)}$$

holds for every tautological elementary implication.

Now $\underline{a} \in A_{(w:G)}$ may be any solution. Then we have to show that

$$\underline{a} \in A_{(w:t_1 \overset{e}{=} t_2)} \cdot$$

The existence of an existence equation $(w: q_1 \overset{e}{=} q_2)$ with

$$(w: G \longrightarrow q_1 \overset{e}{=} q_2) \in IMP_1(\mathcal{A})$$

for every $z \in Z$ so that $z\underline{t}$ is a subterm of q_1 or q_2 implies $z\underline{t} \in \operatorname{dom} \tilde{\underline{a}}$ for every $z \in Z$. Hence, we can build the assignment $\underline{t} \tilde{\underline{a}} \in A_u$. For this assignment and for every $r \in \operatorname{dom}(\widetilde{\underline{t}\tilde{\underline{a}}})$ hold

$$\operatorname{dom}(\widetilde{\underline{t}\tilde{\underline{a}}}) = (\operatorname{dom} \tilde{\underline{a}}) \tilde{\underline{t}}^{-1} \quad \text{and} \quad r(\widetilde{\underline{t}\tilde{\underline{a}}}) = (r\tilde{\underline{t}}) \tilde{\underline{a}} \ .$$

For every $(u: p_1 \overset{e}{=} p_2) \in (u:H)$ we obtain

$$p_1 \tilde{\underline{t}} \in \operatorname{dom} \tilde{\underline{a}}, \quad p_2 \tilde{\underline{t}} \in \operatorname{dom} \tilde{\underline{a}}, \text{ and } \quad (p_1 \tilde{\underline{t}}) \tilde{\underline{a}} = (p_2 \tilde{\underline{t}}) \tilde{\underline{a}}$$

because

$$(w: G \longrightarrow p_1 \tilde{\underline{t}} \overset{e}{=} p_2 \tilde{\underline{t}}) \in IMP_1(\mathcal{A}).$$

This proves $\underline{t} \tilde{\underline{a}} \in A_{(u:H)}$.

$(u: H \longrightarrow r_1 \overset{e}{=} r_2) \in TAUT(\Sigma) \cup \mathcal{A}$ guarantees $\underline{t}\tilde{\underline{a}} \in A_{(u:r_1 \overset{e}{=} r_2)}$, and so

$$r_1, r_2 \in \operatorname{dom}(\widetilde{\underline{t}\tilde{\underline{a}}}), \quad r_1(\widetilde{\underline{t}\tilde{\underline{a}}}) = r_2(\widetilde{\underline{t}\tilde{\underline{a}}}) \ .$$

This together with $t_1 = r_1 \tilde{\underline{t}}$, $t_2 = r_2 \tilde{\underline{t}}$ yields

$$t_1 \tilde{\underline{a}} = (r_1 \tilde{\underline{t}}) \tilde{\underline{a}} = r_1(\widetilde{\underline{t}\tilde{\underline{a}}}) = r_2(\widetilde{\underline{t}\tilde{\underline{a}}}) = (r_2 \tilde{\underline{t}}) \tilde{\underline{a}} = t_2 \tilde{\underline{a}}$$

or in other words

$$\underline{a} \in A_{(w:t_1 \overset{e}{=} t_2)} \ .$$

Finally we extend the proof from elementary implications to arbitrary implications. Therefore let $(v: G \longrightarrow H)$ with $v: X \longrightarrow S$ be any implication with $\mathcal{A} \vdash (v: G \longrightarrow H)$, and A may be any $\mathcal{A}$-algebra. We have to prove that

$$\underline{a} \in A_{(v:t_1 \overset{e}{=} t_2)} \quad \text{results from} \quad \underline{a} \in A_{(v:G)}$$

for every $(v: t_1 \overset{e}{=} t_2) \in (v:H)$.

Corresponding to Definition 3.1.7 there is a finite S-sorted subsystem $w: Y \longrightarrow S$ of $v: X \longrightarrow S$ and a finite subset $G^* \subseteq G$ with

$$w \succ (v: G^*), \quad w \succ (v: t_1 \overset{e}{=} t_2), \quad \text{and}$$

$$\mathcal{A} \vdash (v{\downarrow}w: G^* \longrightarrow t_1 \overset{e}{=} t_2) \ .$$

The restriction of the assignment $\underline{a} \in A_W$ to the finite S-sorted subsystem $w: Y \longrightarrow S$ will be denoted by $\underline{a}{\downarrow}w \in A_w$.
The covering $w \succ (v: t_1 \overset{e}{=} t_2)$ guarantees that

$$\underline{a}{\downarrow}w \in A_{(v\,w: t_1 \overset{e}{=} t_2)} \quad \text{implies} \quad \underline{a} \in A_{(v: t_1 \overset{e}{=} t_2)} .$$

Now let $\underline{a} \in A_{(v:G)}$. Then $\underline{a} \in A_{(v:G^*)}$ because $G^* \subseteq G$.
The covering $w \succ (v:G^*)$ leads to $\underline{a}{\downarrow}w \in A_{(v{\downarrow}w:G^*)}$, and the elementary implication $(v{\downarrow}w: G^* \longrightarrow t_1 \overset{e}{=} t_2)$ with

$$\mathcal{O} \vdash (v{\downarrow}w: G^* \longrightarrow t_1 \overset{e}{=} t_2)$$

yields

$$\underline{a}{\downarrow}w \in A_{(v{\downarrow}w:G^*)} \subseteq A_{(v{\downarrow}w: t_1 \overset{e}{=} t_2)}$$

and so

$$\underline{a} \in A_{(v: t_1 \overset{e}{=} t_2)} .$$

■

3.2 Free partial algebras

In this section we present a useful construction, namely, the construction of an $\mathcal{O}$-algebra $F(\mathcal{O},G,v)$ freely generated by a set of existence equations $(v:G)$. Using this construction we can prove the completeness of $\mathcal{O}$-derivability, the existence of free and relatively free $\mathcal{O}$-algebras, and finally we can give different quotient constructions.

Definition 3.2.1. Let $\Sigma = (S, \alpha: \Omega \longrightarrow S^* \times S)$ be any signature, $\mathcal{O}$ any set of elementary implications, $v: X \longrightarrow S$ any S-sorted system of variables, and let $(v: G)$ be any set of existence equations. A partial algebra $A \in ALG(\Sigma)$ is referred to as $\mathcal{O}$-algebra freely generated by $(v:G)$ if

(1) A satisfies all axioms of $\mathcal{O}$, i.e. $A \in ALG(\mathcal{O})$;
(2) There exists a solution

$$\underline{e} \in A_{(v:G)}$$

such that for every $\mathcal{O}$-algebra B and every solution

$$\underline{b} \in B_{(v:G)}$$

there exists exactly one homomorphism $b^{\#}: A \longrightarrow B$ with $\underline{b} = \underline{e}\, b^{\#}_v$.

■

Evidently, any two $\mathfrak{A}$-algebras A, B freely generated by one and the same set of existence equations (v: G) are isomorphic. On account of this uniqueness up to isomorphism we denote an $\mathfrak{A}$-algebra freely generated by (v: G) by $F(\mathfrak{A},G,v)$.

In a similar way as free total algebras can be constructed from sets of terms we can construct $F(\mathfrak{A},G,v)$ also from sets of terms.

Let us start with the construction of a partial Σ-algebra $F(\mathfrak{A},G,v)$:
For every $s \in S$ we set

$$F^*(\mathfrak{A},G,v)_s = \{(t,G) \mid t \in T(\Sigma,v)_s \text{ and } \mathfrak{A} \vdash (v: G \longrightarrow t \overset{e}{=} t)\}.$$

The next step is the construction of an S-equivalence

$$\equiv \; = (\equiv_s \mid s \in S)$$

in the S-set $F^*(\mathfrak{A},G,v)$ by

$$(t_1,G) \equiv_s (t_2,G) \quad \text{iff} \quad \mathfrak{A} \vdash (v: G \longrightarrow t_1 \overset{e}{=} t_2).$$

It is easy to see that for every $s \in S$ the binary relation $\equiv_s$ in $F^*(\mathfrak{A},G,v)$ is reflexive, symmetric and transitive. So we can define
(3.2.2.a)

$$F(\mathfrak{A},G,v) = F^*(\mathfrak{A},G,v)/\equiv_s \quad \text{for every } s \in S,$$

and denote the elements of $F(\mathfrak{A},G,v)$, i.e. the equivalence classes, by
(3.2.2.b)

$$[t,G] = \{(r,G) \mid \mathfrak{A} \vdash (v: G \longrightarrow t \overset{e}{=} r)\}.$$

n-tuples of elements of $F(\mathfrak{A},G,v)$ will be abbreviated in the following way:
(3.2.2.c)

$$[\underline{t},G] = ([1\underline{t},G],[2\underline{t},G],\dots,[n\underline{t},G]) \in F(\mathfrak{A},G,v)_w$$

where $w: [n] \longrightarrow S$ and $\underline{t} \in T(\Sigma,v)_w$.

Additionally to the S-set $F(\mathfrak{A},G,v)$ we define an Ω-indexed family of fundamental operations in the following way:
If $\sigma \in \Omega$, $\sigma: w \longrightarrow s$, $w: [n] \longrightarrow S$ then the domain of the fundamental operation

$$\sigma^{F(\mathfrak{A},G,v)}$$

is defined by (3.2.2.d)

$$\mathrm{dom}\,\sigma^{F(\mathfrak{A},G,v)} = \{[\underline{t},G] \mid \underline{t} \in T(\Sigma,v)_w \text{ and}$$
$$\mathfrak{A} \vdash (v: G \longrightarrow \underline{t}\sigma^{T(\Sigma,v)} \overset{e}{=} \underline{t}\sigma^{T(\Sigma,v)})\}$$

And the application of $\sigma^{F(\mathcal{O},G,v)}$ is defined by

(3.2.2.e)

$$[\underline{t},G]\,\sigma^{F(\mathcal{O},G,v)} = [\underline{t}\,\sigma^{T(\mathcal{O},v)},G] \;.$$

The independence of the definitions (3.2.2.d) and (3.2.2.e) from the choice of representatives results immediately from the generalized substitution rule (3.1.5).

Before proving that $F(\mathcal{O},G,v)$ is an $\mathcal{O}$-algebra freely generated by $(v:G)$ we give another important property of $F(\mathcal{O},G,v)$.

Theorem 3.2.3. Let $(w:H)$ with $w: Y \longrightarrow S$ be any set of existence equations, where Y may be infinite. An assignment

$$\underline{f} \in F(\mathcal{O},G,v)_w \;,\text{ given by } y\underline{f} = [y\underline{t},G] \;,\; \underline{t} \in T(\Sigma,v)_w \text{ for}$$

every $y \in Y$, is a solution of $(w:H)$ in $F(\mathcal{O},G,v)$ if and only if

$$\mathcal{O} \vdash (v: G \longrightarrow H\tilde{\underline{t}}) \;.$$

Proof. The case $(w:H) = (w:\emptyset)$ is trivial because

$$\mathcal{O} \vdash (v: G \longrightarrow \emptyset).$$

Now let $\underline{f} \in F(\mathcal{O},G,v)_{(w:H)}$. Then

$$\underline{f} \in \operatorname{dom} t_1^{F(\mathcal{O},G,v)} \cap \operatorname{dom} t_2^{F(\mathcal{O},G,v)}$$

and

$$\underline{f}\, t_1^{F(\mathcal{O},G,v)} = \underline{f}\, t_2^{F(\mathcal{O},G,v)}$$

hold for every $(w: t_1 \overset{e}{=} t_2) \in (w:H)$.

In addition one can derive from (3.2.2.d) and (3.2.2.e) by induction the following equalities:

$$\operatorname{dom} t^{F(\mathcal{O},G,v)} = \{[\underline{t},G] \mid \underline{t} \in T(\Sigma,v)_u \,,\, \mathcal{O} \vdash (v: G \longrightarrow t\tilde{\underline{t}} \overset{e}{=} t\tilde{\underline{t}})\} \;,$$

$$[\underline{t},G]\; t^{F(\mathcal{O},G,v)} = [t\tilde{\underline{t}},G]$$

for every $t \in T(\Sigma,u)$, $u: Z \longrightarrow S$.

Consequently, we obtain

$$\mathcal{O} \vdash (v: G \longrightarrow t_1\tilde{\underline{t}} \overset{e}{=} t_1\tilde{\underline{t}}) \,,\, \mathcal{O} \vdash (v: G \longrightarrow t_2\tilde{\underline{t}} \overset{e}{=} t_2\tilde{\underline{t}}) \,,\text{ and}$$

$$[t_1\tilde{\underline{t}},G] = [t_2\tilde{\underline{t}},G]$$

for $\underline{f} = [\underline{t},G] \in F(\mathcal{O},G,v)_{(w:H)}$, and so

$$\mathcal{O} \vdash (v: G \longrightarrow t_1\tilde{\underline{t}} \overset{e}{=} t_2\tilde{\underline{t}})$$

for every $(w: t_1 \overset{e}{=} t_2) \in (w:H)$. This is identical with

$$\mathcal{O} \vdash (v: G \longrightarrow H\tilde{\underline{t}}) \;.$$

For the proof of the converse direction we assume

$$\mathfrak{A} \vdash (v\colon G \longrightarrow H\underline{\tilde{t}})$$

or

$$\mathfrak{A} \vdash (v\colon G \longrightarrow t_1\underline{\tilde{t}} \overset{e}{=} t_2\underline{\tilde{t}})$$

for every $(w\colon t_1 \overset{e}{=} t_2) \in (w\colon H)$.

As a result of this assumption and Corollary 3.1.4

$$\mathfrak{A} \vdash (v\colon G \longrightarrow t_1\underline{\tilde{t}} \overset{e}{=} t_1\underline{\tilde{t}}) \;,\; \mathfrak{A} \vdash (v\colon G \longrightarrow t_2\underline{\tilde{t}} \overset{e}{=} t_2\underline{\tilde{t}})$$

follow, and so we obtain

$$[\underline{t},G] \in \operatorname{dom} t_1^{F(\mathfrak{A},G,v)} \cap \operatorname{dom} t_2^{F(\mathfrak{A},G,v)} \quad \text{and}$$

$$[\underline{t},G] \in F(\mathfrak{A},G,v)_{(w:t_1 \overset{e}{=} t_2)}$$

for every $(w\colon t_1 \overset{e}{=} t_2) \in (w\colon H)$, i.e.

$$\underline{f} = [\underline{t},G] \in F(\mathfrak{A},G,v)_{(w:H)} . \qquad \blacksquare$$

<u>Theorem 3.2.4</u>. For every set of elementary implications $\mathfrak{A}$ and every set of existence equations $(v\colon G)$, $v\colon X \longrightarrow S$, the partial algebra $F(\mathfrak{A},G,v)$ is an $\mathfrak{A}$-algebra freely generated by $(v\colon G)$.

<u>Proof</u>. First of all we prove that $F(\mathfrak{A},G,v)$ is an $\mathfrak{A}$-algebra. Therefore let $(w\colon H \longrightarrow t_1 \overset{e}{=} t_2)$ with $w\colon Y \longrightarrow S$ be any axiom of $\mathfrak{A}$ and $\underline{f} \in F(\mathfrak{A},G,v)_{(w:H)}$. According to Theorem 3.2.3 this means

$$\mathfrak{A} \vdash (v\colon G \longrightarrow H\underline{\tilde{t}})$$

where $\underline{t} \in T(\Sigma,v)_w$ with $\underline{f} = [\underline{t},G]$.

In addition, $\mathfrak{A} \vdash (w\colon H \longrightarrow t_1 \overset{e}{=} t_2)$ holds as a consequence of (3.1.3), so that Corollary 3.1.8 yields

$$\mathfrak{A} \vdash (w\colon G \longrightarrow t_1\underline{\tilde{t}} \overset{e}{=} t_2\underline{\tilde{t}}) .$$

A further application of (3.2.3) leads to

$$\underline{f} = [\underline{t},G] \in F(\mathfrak{A},G,v)_{(w:t_1 \overset{e}{=} t_2)} .$$

Consequently, $F(\mathfrak{A},G,v)$ satisfies every axiom of $\mathfrak{A}$.

Next we show that the assignment

$$\underline{e} \in F(\mathfrak{A},G,v)_v \text{ ,given by } x\underline{e} = [x,G] \text{ for every } x \in X ,$$

is a solution of $(v\colon G)$ in $F(\mathfrak{A},G,v)$.

For $\underline{i} \in T(\Sigma,v)_v$ with $x\underline{i} = x$ for every $x \in X$ the following holds

$$\underline{e} = [\underline{i},G] \text{ and } G = G\underline{\tilde{i}} .$$

This implies

$$\mathcal{O} \vdash (v\colon G \longrightarrow G)$$

so that Theorem 3.2.3 yields

$$\underline{e} \in F(\mathcal{O},G,v)_{(v:G)} \, .$$

Finally we assume that A is any $\mathcal{O}$-algebra and $\underline{a} \in A_{(v:G)}$ any solution. According to (3.2.2.b)

$$\mathcal{O} \vdash (v\colon G \longrightarrow t \overset{e}{=} t)$$

holds for every $[t,G] \in F(\mathcal{O},G,v)$, and by Theorem 3.1.9 we obtain

$$A_{(v:G)} \subseteq A_{(v:t \overset{e}{=} t)} \, .$$

Hence, for every $\underline{a} \in A_{(v:G)}$ it follows that $\underline{a} \in \operatorname{dom} t^A$, and so we can define the homomorphism

$$a^{\#} : F(\mathcal{O},G,v) \longrightarrow A$$

by $[t,G]\, a^{\#} = \underline{a}\, t^A$ for every $[t,G] \in F(\mathcal{O},G,v)$.
It remains to show that this defines really a homomorphism with $\underline{e}\, a^{\#}_v = \underline{a}$.
For every $x \in X$,

$$x\underline{e}\, a^{\#}_v = (x\underline{e})\, a^{\#}_v = [x,G]\, a^{\#}_v = \underline{a}\, x^A = x\underline{a} \, .$$

Now let $\sigma \in \Omega$, with $\sigma : w \longrightarrow s$, $w:[n] \longrightarrow S$, be any operator and

$$[\underline{t},G] \in \operatorname{dom} \sigma^{F(\mathcal{O},G,v)} \, .$$

On account of (3.2.2.d) this implies

$$\mathcal{O} \vdash (v\colon G \longrightarrow \underline{t}\,\sigma^{T(\Sigma,v)} \overset{e}{=} \underline{t}\,\sigma^{T(\Sigma,v)}),$$

and therefore

$$([\underline{t},G]\sigma^{F(\mathcal{O},G,v)})\, a^{\#}_s = [\underline{t}\,\sigma^{T(\Sigma,v)},G]\, a^{\#}_s = \underline{a}(\underline{t}\,\sigma^{T(\Sigma,v)})^A =$$

$$= (\underline{a}(1\underline{t})^A, \ldots, \underline{a}(n\underline{t})^A)\, \sigma^A = (\underline{t}\, a^{\#}_w)\sigma^A.$$

This proves that $a^{\#}\colon F(\mathcal{O},G,v) \longrightarrow A$ is a homomorphism with the required property.
Since the S-set $X^{*} = (X^{*}_s \mid s \in S)$ given by $X^{*}_s = \{[x,G] \mid xv=s\}$ is a generating S-set of $F(\mathcal{O},G,v)$, the assignment $\underline{a} \in A_v$ can be extended to a homomorphism in at most one way. ∎

By means of the preceding results it is easy to prove the completeness of $\mathcal{O}$-derivability.

<u>Theorem 3.2.5</u>. For any two sets of existence equations $(v\colon G)$, $(v\colon H)$,

$$\mathcal{O} \vdash (v\colon G \longrightarrow H)$$

holds if and only if $A_{(v:G)} \subseteq A_{(v:H)}$ is true for every $\mathcal{A}$-algebra A.

Proof. Since Theorem 3.1.9 represents the necessity, we have only to prove the sufficiency.

If $A_{(v:G)} \subseteq A_{(v:H)}$ is true for every $\mathcal{A}$-algebra, then we obtain for $F(\mathcal{A},G,v)$ the relation

$$F(\mathcal{A},G,v)_{(v:G)} \subseteq F(\mathcal{A},G,v)_{(v:H)} \ .$$

For $\underline{i} \in T(\Sigma,v)_v$ with $x\underline{i} = x$ for every $x \in X$,

$$[\underline{i},G] \in F(\mathcal{A},G,v)_{(v:H)}$$

holds, if and only if

$$\mathcal{A} \vdash (v\colon G \longrightarrow H\tilde{\underline{i}}) \ .$$

Since $\underline{i} = Id_{T(\Sigma,v)}$ and

$$[\underline{i},G] \quad F(\mathcal{A},G,v)_{(v:G)} \subseteq F(\mathcal{A},G,v)_{(v:H)}$$

we have proved

$$\mathcal{A} \vdash (v\colon G \longrightarrow H) \ .$$ ■

We illustrate the introduced concept of an $\mathcal{A}$-algebra freely generated by a set of existence equations by applying it to the description of partial algebras. For this reason we recall how laborious the description of two Θ_{CAT}-equoids was in Section 2.4. In Section 2.5 we have shown that the class of small categories can be defined by the signature Σ_{CAT} and by the set of elementary implications $\mathcal{A}^*_{CAT}$. Now we are in a position to define the Θ_{CAT}-equoids A, B of Section 2.4 as $\mathcal{A}^*_{CAT}$-algebras freely generated by suitable sets of existence equations.

The small category A of Section 2.4 is isomorphic to

$$F(\mathcal{A}^*_{CAT},\emptyset,v)$$

where $v\colon\{x,y\} \longrightarrow \{Obj,Mor\}$ with $xv = yv = Mor$.

The small category B of Section 2.4 is isomorphic to

$$F(\mathcal{A}^*_{CAT},G_1,v)$$

with $(v\colon G_1) = (v\colon\{target(x) \overset{e}{=} source(y)\})$.

The same small category B would be described by

$$F(\mathcal{A}^*_{CAT},G_2,w)$$

where $w: \{x,y,z\} \longrightarrow \{Obj, Mor\}$ with $xw = yw = zw = Mor$ and

$$(w: G_2) = (w: \{target(x) \stackrel{e}{=} source(y)\ ,\ z \stackrel{e}{=} comp(x,y)\})\ .$$

If we consider the set of existence equations

$$(w: G_3) = (w: \{target(x) \stackrel{e}{=} source(y)\ ,\ source(x) \stackrel{e}{=} source(z)\ ,\ target(x) \stackrel{e}{=} target(z)\})$$

then $F(\mathfrak{A}^*_{CAT}, G_3, w)$ can be represented by the following diagram:

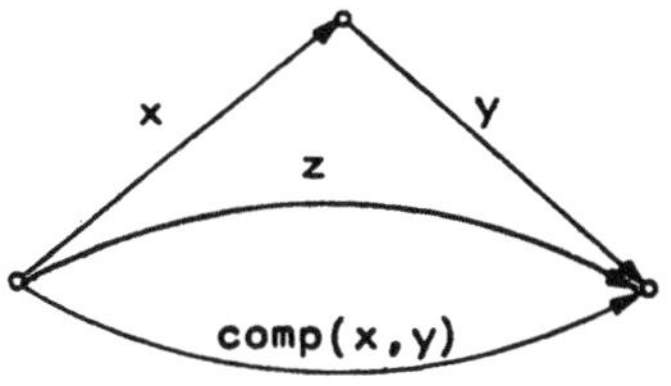

The small category, represented by the following diagram

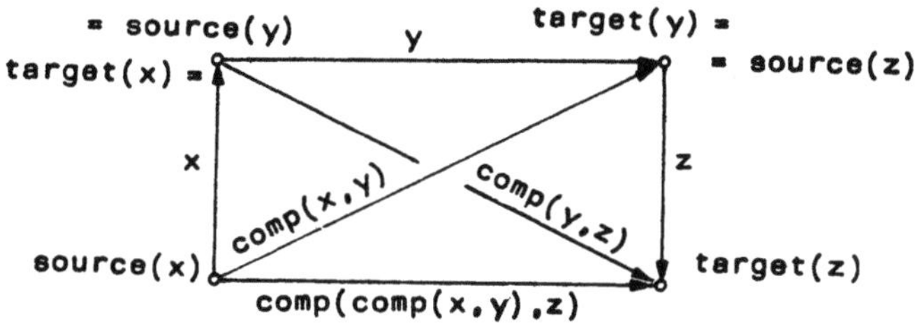

can be described by

$$(w: G_4) = (w: \{target(x) \stackrel{e}{=} source(y)\ ,\ target(y) \stackrel{e}{=} source(z)\})\ .$$

Finally, $F(\mathfrak{A}^*_{CAT}, \emptyset, w)$ describes the following small category:

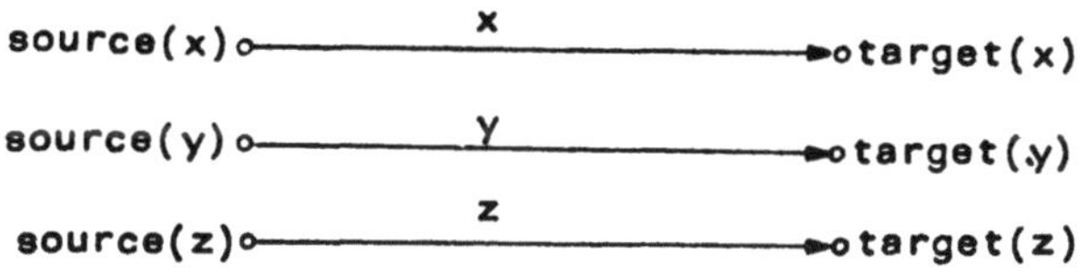

We see that free $\mathfrak{A}^*_{CAT}$-algebras ,i.e. $\mathfrak{A}^*_{CAT}$-algebras freely generated by an empty set of existence equations $(v: \emptyset)$, are of no interest. They coincide with small categories of isolated morphisms.

3.3 Partial quotient algebras

At the end of Section 2.4 we said that the construction of quotient algebras by means of congruence relations cannot be generalized in a straightforward way from total algebras to equoids. We demonstrate this assertion by small categories.

For this reason we consider again the small category A isomorphic to $F(\mathcal{O}\!\mathcal{L}^{*}_{CAT},\emptyset,v)$ with $v:\{x,y\} \longrightarrow \{Obj,Mor\}$ and $xv = yv = Mor$. The two-sorted relation ϱ with

$$\varrho_{Obj} = \{(e,e) \mid e \in A_{Obj}\} \cup \{(target(x),source(y))\} \text{ and}$$
$$\varrho_{Mor} = \{(m,m) \mid m \in A_{Mor}\}$$

is a congruence relation, i.e. ϱ is a two-sorted equivalence relation and the carrier of a subequoid of $A \times A$.
However, the algebraic structure of A cannot be carried over canonically to the two-sorted quotient set A/ϱ, given by

$$(A/\varrho)_{Mor} = \{[x]_\varrho, [y]_\varrho\}$$
$$(A/\varrho)_{Obj} = \{[source(x)]_\varrho, [target(x)]_\varrho = [source(y)]_\varrho, [target(y)]_\varrho\}.$$

Only the unary total operations

$$source^A: A_{Mor} \longrightarrow A_{Obj}, \quad target^A: A_{Mor} \longrightarrow A_{Obj}$$

can be carried over to A/ϱ by

$$source^{A/\varrho}([x]_\varrho) = [source(x)]_\varrho,$$
$$target^{A/\varrho}([x]_\varrho) = [target(x)]_\varrho = [source(y)]_\varrho = source^{A/\varrho}([y]_\varrho)$$
$$target^{A/\varrho}([y]_\varrho) = [target(y)]_\varrho.$$

But now there is a non-trivial solution of the domain condition of the binary composition, namely $([x]_\varrho,[y]_\varrho)$, and there is no congruence class that could represent the value of the composition

$$comp^{A/\varrho}([x]_\varrho,[y]_\varrho).$$

This fact could cause the conjecture that no quotient $\mathcal{O}\!\mathcal{L}^{*}_{CAT}$-algebra A/ϱ exists. But, this conjecture is caused only by lack of success of an elementwise construction. On the other hand the homomorphism $f: A \longrightarrow B$ as defined in Section 2.4 is natural for ϱ.

In the following we show that for every many-sorted relation ϱ, for every set of elementary implications $\mathcal{O}\!\mathcal{L}$, and for every $\mathcal{O}\!\mathcal{L}$-algebra A a homomorphism $f: A \longrightarrow B$ between $\mathcal{O}\!\mathcal{L}$-algebras exists which is natural for ϱ.

To begin with, we return to the preceding example of the homomorphism $f: A \longrightarrow B$ natural for ϱ as defined in Section 2.4 on the pages 94 and 95. We recall that A is isomorphic to $F(\mathcal{O}\!\mathcal{L}^{*}_{CAT}, \emptyset, v)$ and B is isomorphic to

$$F(\mathcal{O}\!\mathcal{L}^{*}_{CAT}, \{target(x) \stackrel{e}{=} source(y)\}, v) \ .$$

We can see that up to isomorphism B is constructed from A by adding the ordered pairs of ϱ, considered as existence equations, to the set of existence equations which describes A. In this construction no missing elements annoy, because

$$[comp(x,y), (v:\{target(x) \stackrel{e}{=} source(y)\})]$$

exists in

$$F(\mathcal{O}\!\mathcal{L}^{*}_{CAT}, \{target(x) \stackrel{e}{=} source(y)\}, v)$$

due to

$$\mathcal{O}\!\mathcal{L}^{*}_{CAT} \vdash (v: \{target(x) \stackrel{e}{=} source(y)\} \longrightarrow comp(x,y) \stackrel{e}{=} comp(x,y)).$$

In order to investigate the general construction of quotient $\mathcal{O}\!\mathcal{L}$-algebras, we show first that every $\mathcal{O}\!\mathcal{L}$-algebra A is isomorphic to an algebra $F(\mathcal{O}\!\mathcal{L}, G, v)$ where $(v: G)$ can be derived canonically from A.

Let $\mathcal{O}\!\mathcal{L}$ be any set of elementary implications and A any $\mathcal{O}\!\mathcal{L}$-algebra of type $\Sigma = (S, \alpha: \Omega \longrightarrow S^{*} \times S)$. Then we define:

$$X_A = \{(a,s) \mid s \in S,\ a \in A_s\} \ ;$$

$$v_A : X_A \longrightarrow S \quad \text{with} \quad (a,s)\, v_A = s \quad \text{for every} \quad (a,s) \in X_A \ ;$$

$$\underline{e}_A \in A_{(v_A)} \quad \text{with} \quad (a,s)\, \underline{e}_A = a \quad \text{for every} \quad (a,s) \in X_A \ .$$

Additionally we set

$$G_A = \{ (v_A : (a,s) \stackrel{e}{=} t) \mid t \in T(\Sigma, v_A)_s,\ s \in S,\ a = \underline{e}_A t^A \} \ .$$

Then $\underline{e}_A$ is a solution of (v_A, G_A) in A, and the unique extension to a homomorphism

$$e^{\#}_A : F(\mathcal{O}\!\mathcal{L}, G_A, v_A) \longrightarrow A$$

is an isomorphism, i.e.

(3.3.1)

$$F(\mathcal{O}\!\mathcal{L}, G_A, v_A) \cong A \ .$$

This representation shows that we do not restrict the generality if we deal only with $\mathfrak{A}$-algebras of the form $F(\mathfrak{A},G,v)$. Obviously, there are many different sets of existence equations which describe one and the same $\mathfrak{A}$-algebra. In general it is recursively undecidable whether $F(\mathfrak{A},G,v)$ and $F(\mathfrak{A},H,w)$ are isomorphic, even in the case that $(v:G)$, $(w:H)$ are finite sets of existence equations on finite S-sorted systems of variables.

Now we can start with the construction of a natural homomorphism

$$f : F(\mathfrak{A},G,v) \longrightarrow F(\mathfrak{A},H,w)$$

for every S-sorted relation ϱ in $F(\mathfrak{A},G,v)$.
For every binary S-relation in $F(\mathfrak{A},G,v)$ we define the so-called associated set of existence equations $(v:G_\varrho)$ by
(3.3.2)

$$(v:G_\varrho) = \{ (v: t_1 \overset{s}{=} t_2) \mid ([t_1,G], [t_2,G]) \ \varrho_s, s \in S \}.$$

<u>Theorem 3.3.3</u>. For every S-sorted binary relation ϱ in $F(\mathfrak{A},G,v)$ the homomorphism

$$f : F(\mathfrak{A},G,v) \longrightarrow F(\mathfrak{A},G \cup G_\varrho,v)$$

with

$$[t,G]f = [t,G \cup G_\varrho] \quad \text{for every} \quad [t,G] \in F(\mathfrak{A},G,v)$$

is natural for ϱ.

<u>Proof</u>. First we show that the previously defined S-function

$$f : F(\mathfrak{A},G,v) \longrightarrow F(\mathfrak{A},G \cup G_\varrho,v)$$

is a homomorphism.
Let $\underline{e} \in F(\mathfrak{A},G,v)_{(v:G)}$ be the universal solution of $(v:G)$, i.e. for every $\mathfrak{A}$-algebra B and every solution $\underline{b} \in B_{(v:G)}$ there exists exactly one homomorphism

$$b^\# : F(\mathfrak{A},G,v) \longrightarrow B \quad \text{with} \quad \underline{e}\, b^\#_v = \underline{b}.$$

The assignment $\underline{1} \in F(\mathfrak{A},G \cup G_\varrho,v)_v$ with $x\underline{1} = [x,G \cup G_\varrho]$ for every $x \in X$ is a solution of $(v:G)$ in $F(\mathfrak{A},G \cup G_\varrho,v)$ because of $G \subseteq G \cup G_\varrho$ and Theorem 3.2.3. The given S-function

$$f : F(\mathfrak{A},G,v) \longrightarrow F(\mathfrak{A},G \cup G_\varrho,v)$$

is then the unique extension of the solution $\underline{1} \in F(\mathfrak{A},G \cup G_\varrho,v)_{(v:G)}$ to a homomorphism. It remains to prove that

$$f : F(\mathfrak{A},G,v) \longrightarrow F(\mathfrak{A},G \cup G_\varrho,v)$$

satisfies condition (2) of Definition 2.4.8.

If B is any $\mathfrak{A}$-algebra and $g: F(\mathfrak{A},G,v) \longrightarrow B$ is a homomorphism with $[t_1,G]g_s = [t_2,G]g_s$ for every $([t_1,G],[t_2,G]) \in \rho_s$, $s \in S$, then the assignment $\underline{e}\,g_v \in B_v$ is a solution of $(v:G \cup G_\rho)$ in B and $g: F(\mathfrak{A},G,v) \longrightarrow B$ is the unique extension of $\underline{e}\,g_v \in B_v$ to a homomorphism. Since $\underline{e}\,g_v \in B_{(v:G \cup G_\rho)}$ there exists exactly one homomorphism $h: F(\mathfrak{A},G \cup G_\rho,v) \longrightarrow B$ with $\underline{1}\,h_v = \underline{e}\,g_v$. By that

$$f \cdot h : F(\mathfrak{A},G,v) \longrightarrow B$$

is a homomorphism with

$$\underline{e}(f \cdot h)_v = (\underline{e}\,f_v)\,h_v = \underline{1}\,h_v = \underline{e}\,g_v \quad \text{and} \quad g = f \cdot h \ .$$

If $h': F(\mathfrak{A},G \cup G_\rho,v) \longrightarrow B$ is a further homomorphism with $f \cdot h' = g$ then

$$\underline{1}\,h'_v = (\underline{e}\,f_v)\,h'_v = \underline{e}(f_v h'_v) = \underline{e}(f \cdot h')_v = \underline{e}\,g_v = \underline{1}\,h_v$$

and so

$$h = h' \ . \qquad \blacksquare$$

Theorem 3.3.3 shows that the preceding construction of the natural homomorphism $f: A \longrightarrow B$ between $\mathfrak{A}^{*}_{CAT}$-algebras is representative for constructions of quotient $\mathfrak{A}$-algebras.

As next we are concerned with a second useful construction, the so-called *pushout-construction*, taken from the theory of categories. Let be given homomorphisms

$$f: A \longrightarrow B \ , \quad g: A \longrightarrow C$$

between $\mathfrak{A}$-algebras with common domain A. A couple of homomorphisms

$$f^{\#}: C \longrightarrow Q \ , \quad g^{\#}: B \longrightarrow Q$$

forms a *pushout* of $f: A \longrightarrow B$, $g: A \longrightarrow C$ if

(i) $f \cdot g^{\#} = g \cdot f^{\#}$;

(ii) for every couple $f': C \longrightarrow Q'$, $g': B \longrightarrow Q'$ with $f \cdot g' = g \cdot f'$ there exists exactly one homomorphism $h: Q \longrightarrow Q'$ with $g^{\#} \cdot h = g'$ and $f^{\#} \cdot h = f'$.

Evidently, the homomorphisms $f: A \longrightarrow B$, $g: A \longrightarrow C$ determine the 'pushout-$\mathfrak{A}$-algebra Q uniquely up to isomorphism.

Before we will give a general construction of pushout-$\mathfrak{A}$-algebras we show that the pushout generalizes the construction of sums or coproducts of $\mathfrak{A}$-algebras. This construction corresponds to the construction of the free sum of groups in the case of general $\mathfrak{A}$-algebras.

The construction of sums can be derived from the more general pushout construction by means of so-called initial $\mathcal{A}$-algebras.

Definition 3.3.4. Let $\mathcal{K} \in$ ALG($\mathcal{A}$) be any class of $\mathcal{A}$-algebras. An $\mathcal{A}$-algebra A is called i n i t i a l with respect to the class $\mathcal{K}$, or initial $\mathcal{K}$-algebra, if

(i) $A \in \mathcal{K}$;

(ii) For every $B \in \mathcal{K}$ there exists exactly one homomorphism $b: A \longrightarrow B$. ∎

Corollary 3.3.5. For every set of elementary implications $\mathcal{A}$ the $\mathcal{A}$-algebra $F(\mathcal{A}, \emptyset, \lambda)$ is initial with respect to ALG($\mathcal{A}$).

Proof. Let B be any $\mathcal{A}$-algebra. Then there exists exactly one assignment of the empty string $\lambda: \emptyset \longrightarrow S$ in the carrier of the $\mathcal{A}$-algebra B, namely the inclusion of the empty set into B. This unique assignment is the only solution of $(\lambda, \emptyset)$ in B, so that by Theorem 3.2.4 exactly one homomorphism $b: F(\mathcal{A}, \emptyset, \lambda) \longrightarrow B$ exists. ∎

We point out that $F(\mathcal{A}, \emptyset, \lambda)$ may be the empty algebra or that some of its carrier sets may be empty. For instance,

$$F(\mathcal{A}^{*}_{CAT}, \emptyset, \lambda)$$

is the empty category.

Now let $a: F(\mathcal{A}, \emptyset, \lambda) \longrightarrow A$, $b: F(\mathcal{A}, \emptyset, \lambda) \longrightarrow B$ be homomorphisms between $\mathcal{A}$-algebras. According to (3.3.5) these homomorphisms are uniquely determined by $A, B \in$ ALG($\mathcal{A}$). If $f: A \longrightarrow C$, $g: B \longrightarrow C$ are homomorphisms between $\mathcal{A}$-algebras with common codomain, then in any case the compositions $a \cdot f$ and $b \cdot g$ are equal and coincide with the unique homomorphism $c: F(\mathcal{A}, \emptyset, \lambda) \longrightarrow C$.

An element-free description of the sum $A + B$ is given by homomorphisms $i: A \longrightarrow A + B$, $j: B \longrightarrow A + B$ such that for each pair of homomorphisms $f: A \longrightarrow C$, $g: B \longrightarrow C$ there exists exactly one homomorphism $h: A + B \longrightarrow C$ with $i \cdot h = f$ and $j \cdot h = g$.
If we now consider the pushout of the uniquely determined homomorphisms $a: F(\mathcal{A}, \emptyset, \lambda) \longrightarrow A$, $b: F(\mathcal{A}, \emptyset, \lambda) \longrightarrow B$ given by the homomorphisms $a^{\#}: B \longrightarrow Q$, $b^{\#}: A \longrightarrow Q$, then the $\mathcal{A}$-algebras Q and $A + B$ are isomorphic. Therefore the construction of the sum of $\mathcal{A}$-algebras is a special case of the construction of a pushout of homomorphisms between $\mathcal{A}$-algebras.

Now we describe the general pushout construction. For this purpose we introduce a notational convention. If

$$f: F(\mathfrak{A}, G_1, v_1) \longrightarrow F(\mathfrak{A}, G_2, v_2)$$

is a homomorphism and $[t, G_1]$ any element of $F(\mathfrak{A}, G_1, v_1)$, then $t\,f$ denotes any term with $[t, G_1]\,f = [t\,f, G_2]$.

Corollary 3.3.6. Let be given homomorphisms

$$f: F(\mathfrak{A}, G_1, v_1) \longrightarrow F(\mathfrak{A}, G_2, v_2),$$
$$g: F(\mathfrak{A}, G_1, v_1) \longrightarrow F(\mathfrak{A}, G_3, v_3)$$

with $v_1: X \longrightarrow S$, $v_2: Y \longrightarrow S$, $v_3: Z \longrightarrow S$ and $Y \cap Z = \emptyset$.
Then

$$f^{\#}: F(\mathfrak{A}, G_3, v_3) \longrightarrow F(\mathfrak{A}, G, v_2+v_3) \quad \text{and}$$
$$g^{\#}: F(\mathfrak{A}, G_2, v_2) \longrightarrow F(\mathfrak{A}, G, v_2+v_3)$$

form a pushout for f and g, where

$$v_2+v_3: Y \cup Z \longrightarrow S \quad \text{with} \quad x(v_2+v_3) = x\,v_2 \quad \text{if} \quad x \in Y \quad \text{and}$$
$$x(v_2+v_3) = x\,v_3 \quad \text{if} \quad x \in Z,$$

and where

$$G = G_2 \cup G_3 \cup \{(v_2+v_3: x\,f \overset{e}{=} x\,g) \mid x \in X\}.$$

The homomorphisms $f^{\#}$ and $g^{\#}$ are defined by

$$[t, G_2]\,g^{\#} = [t, G] \quad \text{for every} \quad [t, G_2] \ F(\mathfrak{A}, G_2, v_2) \quad \text{and}$$
$$[t, G_3]\,f^{\#} = [t, G] \quad \text{for every} \quad [t, G_3] \ F(\mathfrak{A}, G_3, v_3).$$

Proof. $f^{\#}$ and $g^{\#}$ are homomorphisms according to the same reasoning as in the proof of Theorem 3.3.3. So it remains to show that

$$f \cdot g^{\#} = g \cdot f^{\#}$$

and that $f^{\#}$ and $g^{\#}$ form a pushout for f and g.

Let $\underline{1} \in F(\mathfrak{A}, G_1, v_1)_{(v_1 : G_1)}$ denote the universal solution.
If we can prove $\underline{1}(f \cdot g^{\#})_{v_1} = \underline{1}(g\,f^{\#})_{v_1}$ then $f \cdot g^{\#} = g\,f^{\#}$ follows according to condition (2) of Definition 3.2.1
$[x\,f, G] = [x\,g, G]$ holds for every $x \in X$, since

$$(v_2+v_3: x\,f \overset{e}{=} x\,g) \in (v_2+v_3: G) \quad \text{for every} \quad x \in X.$$

This yields

$$x(\underline{1}(f \cdot g^{\#})) = (x\,\underline{1})(f \cdot g^{\#}) = ([x, G_1]\,f)\,g^{\#} = [x\,f, G_2]\,g^{\#} =$$
$$= [x\,f, G] = [x\,g, G] = x(\underline{1}(g \cdot f^{\#})),$$

and so

$$\underline{1}(f \cdot g^{\#}) = \underline{1}(g \circ f^{\#}) \quad \text{and} \quad f \cdot g^{\#} = g \circ f^{\#} .$$

Now let

$$g' : F(\mathcal{O}\!\mathcal{L}, G_2, v_2) \longrightarrow B$$

$$f' : F(\mathcal{O}\!\mathcal{L}, G_3, v_3) \longrightarrow B$$

be homomorphisms with $B \in ALG(\mathcal{O}\!\mathcal{L})$ and $f \cdot g' = g \circ f'$.
Then there is exactly one assignment

$$\underline{b} \in B_{v_2+v_3} \quad \text{with} \quad y\underline{b} = y(\underline{i}\, g') \ \text{for every} \ y \in Y \ \text{and} \quad z\underline{b} = z(\underline{j}\, f') \ \text{for every} \ z \in Z ,$$

where

$$\underline{i} \in F(\mathcal{O}\!\mathcal{L}, G_2, v_2)_{(v_2 : G_2)} , \quad \underline{j} \in F(\mathcal{O}\!\mathcal{L}, G_3, v_3)_{(v_3 : G_3)}$$

are the universal solutions, respectively.

Because of $G_2 \subseteq G$, $G_3 \subseteq G$, $v_2 \leq v_2+v_3$, and $v_3 \leq v_2+v_3$ we can interpret G_2 and G_3 as sets of existence equations on v_2+v_3 , and so we obtain

$$\underline{b} \in B_{(v_2+v_3 : G_2)} \ \text{and} \ \underline{b} \in B_{(v_2+v_3 : G_3)} .$$

Now let $x \in X$, $t_1 \in T(\Sigma, v_2)$, $t_2 \in T(\Sigma, v_3)$ with $x f = [t_1, G_2]$ and $x g = [t_2, G_3]$, then

$$(v_2+v_3 : t_1 \overset{e}{=} t_2) \in G \setminus (G_2 \cup G_3)$$

because $t_1 = x f$ and $t_2 = x g$.
Since $\underline{b}\, t_1^B$ does not depend on $z \in Z$, and $\underline{b}\, t_2^B$ does not depend on $y \in Y$ the following holds:

$$\underline{b}\, t_1^B = (\underline{i}\, g')\, t_1^B = (\underline{i}\, t_1^{F(\mathcal{O}\!\mathcal{L}, G_2, v_2)})\, g' = [t_1, G]\, g' = (x f)\, g' =$$
$$= x(f \cdot g') = x(g \circ f') = [t_2, G_3]\, f' = \underline{b}\, t_2^B .$$

This means

$$\underline{b} \in B_{(v_2+v_3 : G)} .$$

Consequently, there is exactly one homomorphism

$$h : F(\mathcal{O}\!\mathcal{L}, G, v_2+v_3) \longrightarrow B \quad \text{with} \quad \underline{k}\, h_{(v_2+v_3)} = \underline{b} ,$$

where

$$\underline{k} \in F(\mathcal{O}\!\mathcal{L}, G, v_2+v_3)_{(v_2+v_3 : G)}$$

denotes the universal solution.

If $\underline{k}{\downarrow}v_2$, $\underline{k}{\downarrow}v_3$, $\underline{b}{\downarrow}v_2$, $\underline{b}{\downarrow}v_3$ denote the restrictions of $\underline{k}$ and $\underline{b}$ to the S-sorted subsystems v_2 and v_3 of v_2+v_3, respectively, it follows

$$\underline{1}f^{\#} = \underline{k}\ v_3,\ \underline{1}g^{\#} = \underline{k}\,v_2,\ \underline{1}f' = \underline{b}{\downarrow}v_3,\ \underline{1}g' = \underline{b}{\downarrow}v_2$$

and additionally

$$\underline{1}f^{\#}h = (\underline{k}{\downarrow}v_3)\,h = (\underline{k}\,h)\,v_3 = \underline{b}{\downarrow}v_3 = \underline{1}g'$$

$$\underline{1}g^{\#}h = (\underline{k}{\downarrow}v_2)\,h = (\underline{k}\,h)\,v_2 = \underline{b}{\downarrow}v_2 = \underline{1}f',$$

and so

$$f^{\#}\circ h = f' \quad \text{and} \quad g^{\#}\circ h = g'\ .$$ ∎

Since the pushout construction is not so familiar outside the theory of categories we illustrate this construction by some examples. Furthermore, this construction will be used extensively in Chapter 4 for the construction of complex specifications of abstract data types of simpler ones.

First we remark that the pushout construction encloses the 'union of substructures'. Evidently, the set-theoretic union of $\mathfrak{A}$-subalgebras is in general not the carrier of an $\mathfrak{A}$-subalgebra. By the union $A \cup B$ of $\mathfrak{A}$-subalgebras A and B of an $\mathfrak{A}$-algebra C we understand the smallest $\mathfrak{A}$-subalgebra of C containing both A and B. Therefore, the union of $\mathfrak{A}$-subalgebras is the $\mathfrak{A}$-subalgebra generated by the set-theoretic union.

Now let $A \subseteq C$, $B \subseteq C$ be $\mathfrak{A}$-subalgebras of an $\mathfrak{A}$-algebra C, and $i: A \cap B \longrightarrow A$, $j: A \cap B \longrightarrow B$ denote the inclusion of the intersection $A \cap B$ in A and B, respectively. If $i^{\#}: B \longrightarrow Q$ and $j^{\#}: A \longrightarrow Q$ form a pushout of i and j, then Q is isomorphic to the union of A and B, or in other words, the union $A \cup B$ with the inclusions of A and B into the structural union $A \cup B$ form a pushout of i and j.

Roughly speaking, the pushout construction of given homomorphisms $f: A \longrightarrow B$, $g: A \longrightarrow C$ corresponds to some kind of structural sticking together of B and C, where the common plane of intersection is determined by $f: A \longrightarrow B$ and $g: A \longrightarrow C$. Surely, $\mathfrak{A}$-algebras are not sticked together to planes but they are sticked together to $\mathfrak{A}$-algebras.

This interpretation of pushout constructions will be illustrated by means of small categories. Categories are very well for this task, because we can work with finite $\mathfrak{A}$-algebras.

Let be given the following two-sorted systems of variables:

$v_1 : \{x\} \longrightarrow \{Obj, Mor\}$ with $x v_1 = Mor$,

$v_2 : \{y, z\} \longrightarrow \{Obj, Mor\}$ with $y v_2 = z v_2 = Mor$,

$v_3 : \{y^*, z^*\} \longrightarrow \{Obj, Mor\}$ with $y^* v_3 = z^* v_3 = Mor$.

We start with the homomorphisms

$f_1 : F(\mathcal{A}^{*}_{CAT}, \emptyset, v_1) \longrightarrow F(\mathcal{A}^{*}_{CAT}, \{target(y) \overset{e}{=} source(z)\}, v_2)$

given by $x f_1 = y$;

$g_1 : F(\mathcal{A}^{*}_{CAT}, \emptyset, v_1) \longrightarrow F(\mathcal{A}^{*}_{CAT}, \{target(y^*) \overset{e}{=} source(z^*)\}, v_3)$

given by $x g_1 = y^*$.

The pushout-category of f_1 and g_1 is given by

$F(\mathcal{A}^{*}_{CAT}, \{y \overset{e}{=} y^*,\ target(y) \overset{e}{=} source(z), target(y^*) \overset{e}{=} source(z^*)\}, v_2 + v_3)$.

A graphical representation of this pushout construction is the following:

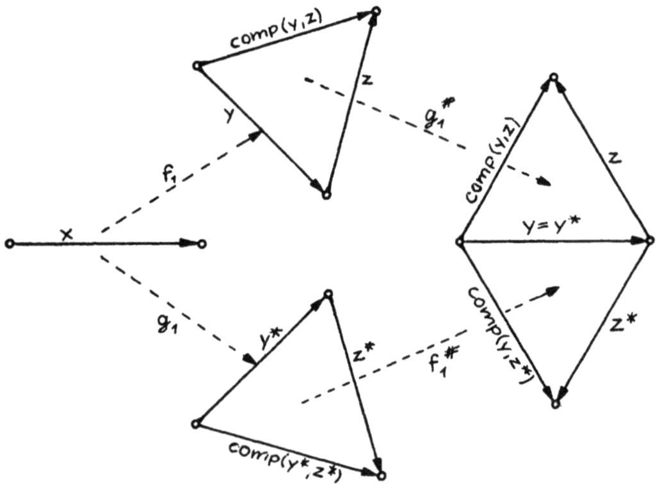

As next we take the homomorphisms

$f_2 : F(\mathcal{A}^{*}_{CAT}, \emptyset, v_1) \longrightarrow F(\mathcal{A}^{*}_{CAT}, \{target(y) \overset{e}{=} source(z)\}, v_2)$

given by $x f_2 = y$

$g_2 : F(\mathcal{A}^{*}_{CAT}, \emptyset, v_1) \longrightarrow F(\mathcal{A}^{*}_{CAT}, \{target(y^*) \overset{e}{=} source(z^*)\}, v_3)$

given by $x g_2 = z^*$.

Now the pushout-category of f_2 and g_2 is given by

$F(\mathcal{A}^{*}_{CAT}, \{z^* \overset{e}{=} y,\ target(y) = source(z), target(y^*) = source(z^*)\}, v_2 + v_3)$

which can be represented graphically as follows:

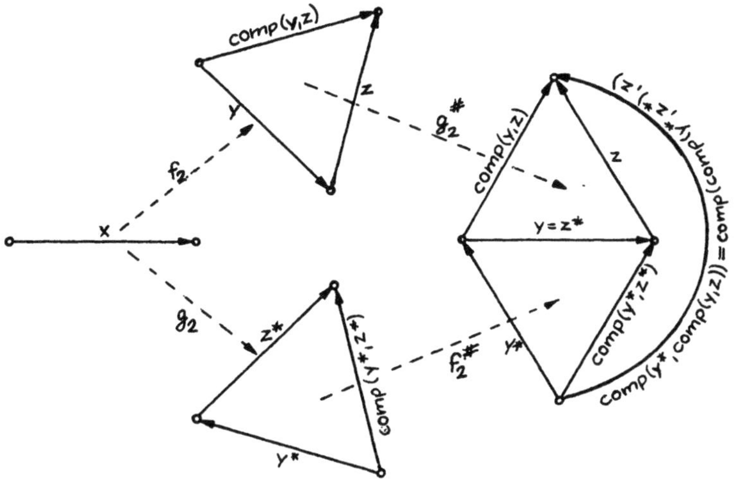

Finally we consider the homomorphisms

$f_3\colon F(\mathcal{A}^{*}_{CAT},\emptyset,v_1) \longrightarrow F(\mathcal{A}^{*}_{CAT},\{target(y)\overset{e}{=}source(z)\},v_2)$

given by $x f_3 = y$,

$g_3\colon F(\mathcal{A}^{*}_{CAT},\emptyset,v_1) \longrightarrow F(\mathcal{A}^{*}_{CAT},\{target(y^*)\overset{e}{=}source(z^*)\},v_3)$

given by $x g_3 = comp(y^*,z^*)$.

The pushout-category of f_3 and g_3 is given by

$$F(\mathcal{A}^{*}_{CAT},\{y\overset{e}{=}comp(y^*,z^*),target(y)\overset{e}{=}source(z),target(y^*)\overset{e}{=}source(z^*)\},\ v_2+v_3) .$$

An illustrating diagram is

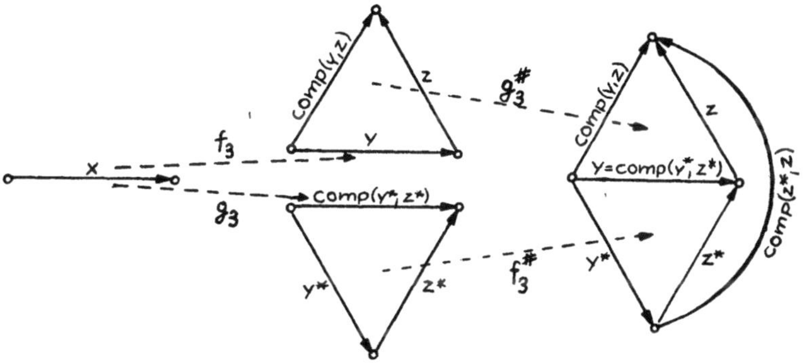

We hope that these examples are helpful for a better understanding of pushout constructions of $\mathcal{A}$-algebras.

3.5 Theorem of Homomorphisms

The quotient construction in Section 3.3 makes visible an essential difference in the behaviour of total algebras and of partial ones. A natural homomorphism between $\mathfrak{A}$-algebras is not necessarily a surjective one, i.e. the application of fundamental operations in a quotient $\mathfrak{A}$-algebra cannot be reduced automatically to applications of fundamental operations to representatives. Does this fact affect the possibility of the reconstruction of the homomorphic image from the domain and the kernel of the homomorphism?

We have defined the notion of a subalgebra and of the homomorphic image in Section 2.4 only for equoids. Therefore we restrict the considerations of this section also to equoids as distinct from the preceding sections of this chapter.

The results of the preceding section can and will be used in this section. But, we have to take into account the fact that now the set of elementary implications $\mathfrak{A}$ is given as well by the necessary and sufficient domain conditions as by axioms interrelating the fundamental operations.

In the case of total algebras the theorem of homomorphisms causes that exactly the surjective homomorphisms are natural for its kernel.

First we investigate this problem for small categories, i.e. for $\mathfrak{A}_{CAT}$-equoids or $\mathfrak{A}^*_{CAT}$-algebras.

Let us consider the surjective homomorphism

$$f: A \longrightarrow B$$

with

$$A = F(\mathfrak{A}^*_{CAT}, \emptyset, v)$$

$$B = F(\mathfrak{A}^*_{CAT}, x_3 \overset{e}{=} comp(x_1,x_2), v)$$

$$v: \{x_1, x_2, x_3\} \longrightarrow \{Obj, Mor\} \quad \text{with} \quad x_i v = Mor \quad \text{for } i=1,2,3,$$

$$x_i f_{Mor} = x_i \quad \text{for } i=1,2,3 .$$

The kernel of this homomorphism is given by

$$(\ker f)_{Mor} = \{ (x_i, x_i) \mid i=1,2,3 \}$$

$$(\ker f)_{Obj} = \{ (source(x_1), source(x_3)), (target(x_2), target(x_3)), (target(x_1), source(x_2)) \} \cup$$
$$\cup \{ (x,x) \mid x \in A_{Obj} \} .$$

We visualize this situation by the following diagram:

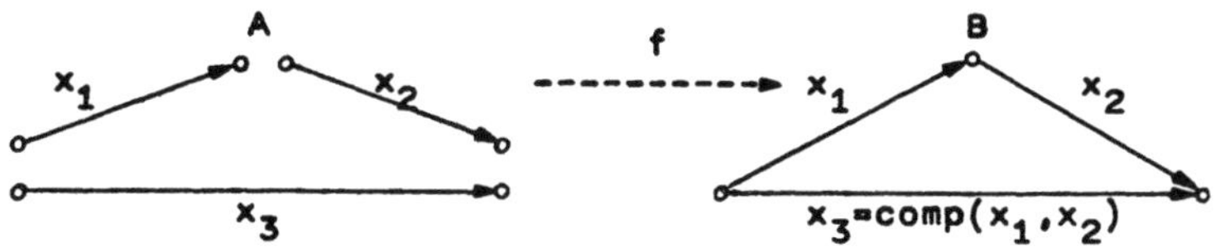

But now $f\colon A \longrightarrow B$ is not natural for its kernel. For the proof of this assertion we consider the following homomorphism

$$g\colon A \longrightarrow C$$

with

$$C = F(\mathcal{A}^{*}_{CAT}, G_1, v)$$

$$G_1 = (v\colon \{\,source(x_1) \stackrel{e}{=} source(x_3),\ target(x_2) \stackrel{e}{=} target(x_3),\ target(x_1) \stackrel{e}{=} source(x_2)\,\}\ )\ , \text{ and}$$

$$x_i\, g = x_i \quad \text{for } i=1,2,3.$$

The small category can be graphically described as follows:

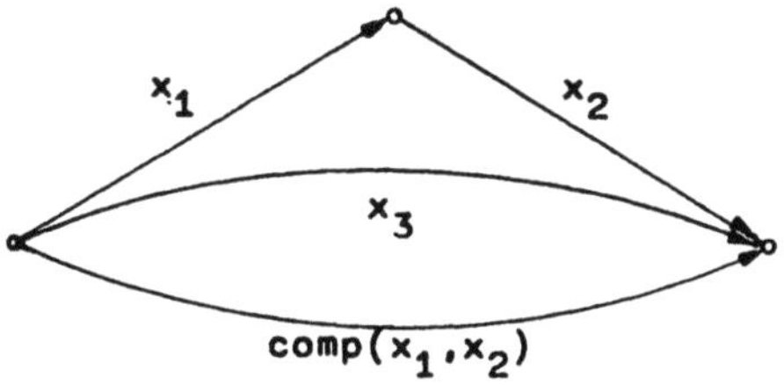

According to Section 3.4 the homomorphism $g\colon A \longrightarrow C$ is natural for its kernel. Since B and C are not isomorphic and $\ker f = \ker g$, the homomorphism $f\colon A \longrightarrow B$ cannot be natural for its kernel.

Let $f\colon A \longrightarrow B$ be any homomorphism between small categories and $q_1\colon A \longrightarrow Q_1$ may be natural for $\ker f$, so that a functor

$$h_1\colon Q_1 \longrightarrow B \quad \text{with} \quad f = q_1 \circ h_1$$

exists.

Q_1 is isomorphic to the homomorphic image $\langle A\, f\rangle$ if and only if $h_1\colon Q_1 \longrightarrow B$ is injective. One can prove generally that Q_1 and $\langle A\, f\rangle$ are not isomorphic iff there are morphisms $x,y,z\ A_{Mor}$ with $x \neq source^A(x)$, $y \neq source^A(y)$, $z \neq source^A(z)$, $target^A(x) \neq source(y)$, and $z\, f_{Mor} = comp^B(x\, f_{Mor}, y\, f_{Mor})$.

If $h_1\colon Q_1 \longrightarrow B$ is not injective there is a non-trivial homomorphism

$$q_2\colon Q_1 \longrightarrow Q_2$$

which is natural for $\ker h_1$ and there is a homomorphism

$h_2: Q_2 \longrightarrow B$ with $h_1 = q_2 \circ h_2$.

One can prove additionally that

$$(h_2)_{Obj} = (Q_2)_{Obj} \longrightarrow B_{Obj}$$

is always injective, and so in Q_1 there are no morphisms x,y,z having the properties described above with respect to $h_1: Q_1 \longrightarrow B$. This implies that $h_2: Q_2 \longrightarrow B$ becomes injective for both sorts Obj, Mor and that the homomorphic image $\langle A\, f\rangle$ becomes isomorphic to Q_2. These considerations show that the homomorphic image $\langle A\, f\rangle$ of a homomorphism between small categories can be reconstructed by one or by two factorizations Q_1 or Q_2, respectively. This existence of iterated quotients motivates the following definition.

Definition 3.4.1. Let Θ be any hep-signature and $\mathfrak{A}$ any set of elementary implications. A finite or countable sequence

$$((f_i, q_i, f_{i+1}) \mid i \in I),$$

of homomorphisms between $\mathfrak{A}$-equoids is called a c h a i n o f i t e r a t e d q u o t i e n t s o f $f: A \longrightarrow B$, where $A, B \in ALG(\mathfrak{A})$, if

(1) $f = f_1$ and $f_i = q_i \circ f_{i+1}$ for every $i \in I$;

(2) $q_i: Q_i \longrightarrow Q_{i+1}$ is not an isomorphism for every $i \in I$;

(3) q_i is natural for $\ker f_i$ for every $i \in I$.

If $I = [n]$, then the natural number n is called the l e n g t h of the chain of iterated quotients of $f: A \longrightarrow B$. If alternatively $I = \mathbb{N}$, then we say that the chain of iterated quotients has infinite length. ∎

With this notion the theorem of homomorphism of total algebras can be formulated as follows: Every chain of iterated quotients of a homomorphism between total algebras is of length one. The preceding considerations show that every chain of iterated quotients of a homomorphism between small categories is at most of length two (see Reichel (1969)). According to the following theorem the length of chains of iterated quotients is not bounded in the general case.

Theorem 3.4.2. There are chains of iterated quotients of homomorphisms between equoids of infinite length.

Proof: Theorem 3.4.2 will be proved by constructing an infinite chain of iterated quotients. According to that let us consider the hep-signature

Θ <u>is</u> <u>sorts</u> N , P

<u>oprn</u> $n \longrightarrow N$

$s(N) \longrightarrow N$

$m(P) \longrightarrow N$

$r(x:N$ <u>iff</u> $xs = n) \longrightarrow P$

<u>end</u> Θ

and let us take the Θ-equoids A and B .
The Θ-equoid A is given by

$A_N = \{(0,x) \mid x \in \mathbb{N}\}$, $A_P = \emptyset$,

$n^A = (0,0)$,

$(0,x)\, s^A = (0,x+1)$ for every $x \in \mathbb{N}$, and

$\mathrm{dom}\, m^A = \mathrm{dom}\, r^A = \emptyset$.

The Θ-equoid B may be defined as follows:

$B_N = \mathbb{N} \times \mathbb{N}$, $B_P = \mathbb{N}$,

$n^B = (0,0)$,

$(x,y)\, s^B = (0,0)$ for every $(x,y) \in B_N$,

$x\, m^B = (x+1,0)$ for every $x \in B_P$,

$\mathrm{dom}\, r^B = B_N$ and $(x,y)\, r^B = x$ for every $(x,y) \in B_N$.

One can see easily that A is an initial Θ-equoid, so that exactly one homomorphism

$f: A \longrightarrow B$

exists, which is given by

$(0,x)\, f_N = (0,0)$ for every $(0,x) \in A_N$,

$f_P: \emptyset \longrightarrow B_P$.

Therefore, the kernel of $f: A \longrightarrow B$ is given by

$(\ker f)_N = A_N \times A_N$ and

$(\ker f)_P = A_P \times A_P$.

By virtue of the construction of natural homomorphisms, according to Theorem 3.3.3, we obtain the first quotient $q_1: A \longrightarrow Q_1$ by setting:

$(Q_1)_N = \{(0,0),(1,0),(1,1),\ldots,(1,x),\ldots \mid x \in \mathbb{N}\}$,

$(Q_1)_P = \{0\}$,

$n^{(Q_1)} = (0,0)$,

$(0,0)\, s^{(Q_1)} = (0,0)$, $(1,x)\, s^{(Q_1)} = (1,x+1)$ for every $x \in \mathbb{N}$,

$0\, m^{(Q_1)} = (1,0)$,

$$\operatorname{dom} r^{(Q_1)} = (0,0) \quad \text{and} \quad (0,0)\, r^{(Q_1)} = 0 .$$

The uniquely determined homomorphism

$$f_2 : Q_1 \longrightarrow B \quad \text{with} \quad f = f_1 = q_1 \circ f_2$$

is given by

$$(j,0)\,(f_2)_N = (j,0) \quad \text{for} \quad j=0,1 ,$$

$$(1,x)\,(f_2)_N = (0,0) \quad \text{for every} \quad x \in \mathbb{N} \quad \text{with} \quad x \geq 1 , \quad \text{and}$$

$$0\,(f_2)_P = 0 .$$

Inductively we define (f_k, q_k, f_{k+1}) for $k \geq 2$ by

$$(A_k)_N = \{(0,0),(1,0),\ldots,(k,0),(k,1),\ldots,(k,x),\ldots \mid x \in \mathbb{N}\} ,$$

$$(A_k)_P = \{0,1,\ldots,k-1\} ,$$

$$n^{(A_k)} = (0,0)$$

$$(j,x)\, s^{(A_k)} = \begin{cases} (0,0) & \text{if} \quad x=0 \quad \text{and} \quad j \leq k-1 \\ (k,x+1) & \text{if} \quad k=j , \quad x \in \mathbb{N} , \end{cases}$$

$$j\, m^{(A_k)} = (j+1,0) \quad \text{for} \quad j=0,1,\ldots,k-1 ,$$

$$\operatorname{dom} r^{(A_k)} = \{(0,0),(1,0),\ldots,(k-1,0)\} ,$$

$$(j,0)\, r^{(A_k)} = j \quad \text{for} \quad j=0,1,\ldots,k-1 .$$

The uniquely determined homomorphism

$$f_k : Q_k \longrightarrow B$$

is given by

$$(j,0)\,(f_k)_N = (j,0) \quad \text{for} \quad j=0,1,\ldots,k ,$$

$$(k,x)\,(f_k)_N = (0,0) \quad \text{for every} \quad x \in \mathbb{N} \quad \text{with} \quad x \neq 0 ,$$

$$j\,(f_k)_P = j \quad \text{for} \quad j=0,1,\ldots,k-1 ,$$

and the homomorphism

$$q_k : Q_k \longrightarrow Q_{k+1}$$

which is natural for $\ker f_k$ is defined by

$$(j,0)\,(q_k)_N = (j,0) \quad \text{for} \quad j=0,1,\ldots,k ,$$

$$(k,x)\,(q_k)_N = (0,0) \quad \text{for every} \quad x \in \mathbb{N} \quad \text{with} \quad x \neq 0 ,$$

$$j\,(q_k)_P = j \quad \text{for} \quad j=0,1,\ldots,k-1 .$$

This explicitly defined infinite chain of iterated quotients of a homomorphism $f : A \longrightarrow B$ proves Theorem 3.4.2. ∎

Now we are looking into significant relations between the maximal chain of iterated quotients of a homomorphism $f: A \longrightarrow B$ and the homomorphic image $\langle A f \rangle$.

For this reason we consider the homomorphic image $\langle A f \rangle$ of the homomorphism $f: A \longrightarrow B$ of Proof 3.4.2. We see that

$$\langle A f \rangle_N = \{(x,0) \mid x \in \mathbb{N}\}, \quad \langle A f \rangle_P = \mathbb{N},$$

so that $\langle A f \rangle$ is an infinite subequoid, whereas the set-theoretic image consists of one element, namely $(0,0)$. Furthermore we see that with increasing index k the kth iterated quotient Q_k becomes more and more similar to the homomorphic image. Hence, the following question arises. Can the homomorphic image $\langle A f \rangle$ of a homomorphism be considered as the limit of the maximal chain of iterated quotients of $f: A \longrightarrow B$?

In both cases of total algebras and small categories, where the maximal chains of iterated quotients are of finite length, this question can be answered positively, because we can evidently define that the last quotient of a finite chain of iterated quotients is the limit of this chain.

But what is the limit of an infinite chain of iterated quotients? The following definition of the limit of a countable chain of morphisms

$$A_1 \xrightarrow{f_1} A_2 \xrightarrow{f_2} A_3 \xrightarrow{f_3} \cdots \xrightarrow{f_{n-1}} A_n \xrightarrow{f_n} \cdots$$

is derived from the theory of categories, where it corresponds to the colimit ot the chain.

<u>Definition 3.4.3</u>. Let Θ be a hep-signature, $\mathcal{A}$ a set of elementary implications, and

$$(f_i : A_i \longrightarrow A_{i+1} \mid i \in \mathbb{N})$$

a countable chain of $\mathcal{A}$-equoids. An $\mathcal{A}$-equoid L is called a l i m i t of this chain, if

(1) there are canonical morphisms

$q_i : A_i \longrightarrow L$ with $q_i = f_i \circ q_{i+1}$ for every $i \in \mathbb{N}$, and

(2) for every family of homomorphisms

$(g_i : A_i \longrightarrow B)$ with $g_i = f_i \circ g_{i+1}$ for every $i \in \mathbb{N}$

there is exactly one homomorphism

$h: L \longrightarrow B$ with $g_i = q_i \circ h$ for every $i \in \mathbb{N}$. ∎

Intuitively, condition (1) requires that L is an upper bound and condition (2) implies that L is a least upper bound of the chain. Additionally, condition (2) guarantees that a limit of a chain is unique up to isomorphism.

If we allow in Definition 3.4.3 finite chains

$$(f_i : A_i \longrightarrow A_{i+1} \mid i \in [n])$$

and modify conditions (1) and (2) for finite index sets, then every finite chain

$$A_1 \xrightarrow{f_1} A_2 \xrightarrow{f_2} \ . \ . \ . \ \xrightarrow{f_n} A_{n+1}$$

has a limit, namely $L = A_{n+1}$ where the canonical homomorphisms

$$q_i : A_i \longrightarrow L \quad \text{for} \quad i \in [n+1]$$

are defined by

$$q_{n+1} = Id_{A_{n+1}}, \quad q_n = f_n, \quad q_{n-1} = f_{n-1} \circ f_n ,$$

$$q_i = f_i \circ f_{i+1} \circ \ldots \circ f_{n-1} \circ f_n , \text{ i.e.}$$

$$q_i = f_i \circ q_{i+1} \quad \text{for} \quad i \in [n] .$$

As next we are concerned with the existence of limits of infinite chains.

Corollary 3.4.4. Let Θ be a hep-signature and $\mathfrak{A}$ any set of elementary implications. For every infinite chain

$$(f_i : A_i \longrightarrow A_{i+1} \mid i \in \mathbb{N})$$

of homomorphisms between $\mathfrak{A}$-equoids there exists a limit $L \in ALG(\Theta, \mathfrak{A})$ which is unique up to isomorphism.

Proof. According to general theorems on colimits in categories (see MacLane (1971)) Corollary 3.4.4 would be an immediate consequence of Theorem 3.3.3 and Corollary 3.3.6. But we will give here an explicit construction of L .

If S is the set of sort names of Θ, we define for every $s \in S$ the set

$$L_s^* = \{ (x,j) \mid j \in \mathbb{N}, \ x \in (A_j)_s \} .$$

For every ordered pair $(i,j) \in \mathbb{N} \times \mathbb{N}$ with $i < j$ we define a homomorphism

$$f_{i,j} : A_i \longrightarrow A_j \quad \text{by} \quad f_{i,j} = f_i \circ f_{i+1} \circ \ldots \circ f_{j-1} .$$

Now we define an S-equivalence $(\equiv_s \mid s \in S)$ on

$$L^* = (L^*_s \mid s \in S)$$

by

$(x,i) \equiv_s (y,j)$ if there is a $k \in \mathbb{N}$ with $i \leq k$, $j \leq k$ and $x f_{i,k} = y f_{j,k}$.

Evidently, this relation is an S-equivalence and so we can construct the quotient sets $L_s = L^*_s/\equiv_s$, where the equivalence classes are denoted by

$$[x,j] = \{ (y,i) \mid (x,j) \equiv_s (y,i) \}.$$

In the second step we define the family of fundamental operations

$$(\sigma^L, \sigma \in \Omega)$$

on the S-set $L = (L_s \mid s \in S)$ in the following way:

Let $\sigma \in \Omega$, $\sigma : w \longrightarrow s$, $w: [n] \longrightarrow S$. Then we set

$$\text{dom}\,\sigma^L = \{ ([x_1,i_1], \ldots, [x_n,i_n]) \mid \text{there is a } k \in \mathbb{N} \text{ and there are } y_1, \ldots, y_n \text{ with } i_1 \leq k, \ldots, i_n \leq k, [x_1,i_1] = [y_1,k], \ldots, [x_n,i_n] = [y_n,k] \text{ and } (y_1, \ldots, y_n) \in \text{dom}\,\sigma^{(A_k)} \}, \text{ and}$$

$$([x_1,i_1], \ldots, [x_n,i_n])\,\sigma^L = [(y_1, \ldots, y_n)\,\sigma^{(A_k)}, k]$$

for every $([x_1,i_1], \ldots, [x_n,i_n]) \in \text{dom}\,\sigma^L$.

Straightforward considerations show that $\text{dom}\,\sigma^L$ and the application of σ^L is defined independently of the choice of representatives, and that L is really an $\mathfrak{A}$-equoid.

The canonical morphisms

$q_i : A_i \longrightarrow L$ are given by $x(q_i)_s = [x,i]$

for every $s \in S$, $x \in (A_i)_s$, $i \in \mathbb{N}$.

If $(g_i : A_i \longrightarrow B \mid i \in \mathbb{N})$ is any other family of homomorphisms with $g_i = f_i \circ g_{i+1}$ for every $i \in \mathbb{N}$, we can define an S-function

$h: L \longrightarrow B$ by $[x,i]\,h_s = x(g_i)_s$

for every $s \in S$, $[x,i] \in L_s$. We show that $h: L \longrightarrow B$ is defined independently from the choice of representatives. If $[x,i] = [y,j] \in L_s$, then there exists a natural number $k \in \mathbb{N}$ with $i \leq k$, $j \leq k$ and $x f_{i,k} = y f_{j,k}$ and so $[x,i]\,h = x g_i = x(f_{i,k} \circ g_k) = (x f_{i,k}) g_k = (y f_{i,k}) g_k = y(f_{j,k} \circ g_k) = y g_j = [y,j]\,h$.

It remains to show that $h: L \longrightarrow B$ is a homomorphism.
Let $\sigma \in \Omega$, $\sigma: w \longrightarrow s$, $w: [n] \longrightarrow S$, and $([x_1,i_1],\dots,[x_n,i_n])$ in $\operatorname{dom} \sigma^L$. According to the definition of $\operatorname{dom} \sigma^L$ we obtain

$$(([x_1,i_1],\dots,[x_n,i_n])\sigma^L)h = [(y_1,\dots,y_n)\sigma^{(A_k)},k]h =$$
$$= ((y_1,\dots,y_n)\sigma^{(A_k)})g_k = (y_1 g_k,\dots,y_n g_k)\sigma^B =$$
$$= ([y_1,k]h,\dots,[y_n,k]h)\sigma^B = ([x_1,i_1]h,\dots,[x_n,i_n]h)\sigma^B .$$

Additionally $x(q_i \circ h) = [x,i]h = x g_i$ holds for every $i \in \mathbb{N}$, i.e.
$q_i \circ h = g_i$ holds for every $i \in \mathbb{N}$.
If, on the other hand, $h^*: L \longrightarrow B$ is a homomorphism with $q_i \circ h^* = g_i$ for every $i \in \mathbb{N}$, then

$$[x,i]h = x g_i = x(q_i \circ h^*) = [x,i]h^*$$

for every $i \in \mathbb{N}$ so that finally $h = h^*$. ∎

We want to remark that the Θ-equoid L is constructed by the set-theoretic limit of the chain of S-functions on which in a canonical way the additional algebraic structure, namely the Ω-indexed family of fundamental operations, is carried over from the elements of the chain. This leads to the fact that the application of any fundamental operation σ^L to any n-tuple of arguments can be reduced to a suitable element of the chain. Locally we can compute in a suitable element of the chain instead of in the limit of the chain. But we cannot say that different results obtained by local computing in an element of the chain represent different results in the limit structure L. The relations between the elements of a chain and the limit structure can very well be studied by the example of Proof 3.4.2. According to the following theorem of homomorphisms the homomorphic image $\langle A f \rangle$ is always the limit of the chain of iterated quotients.

(3.4.5) <u>Theorem of Homomorphisms</u>

Let Θ be any hep-signature, $\mathfrak{A}$ any set of elementary implications, and $f: A \longrightarrow B$ any homomorphism between $\mathfrak{A}$-equoids. The homomorphic image $\langle A f \rangle \subseteq B$ is a limit of every maximal chain of iterated quotients of $f: A \longrightarrow B$.

<u>Proof</u>. If both

$$((f_i,q_i,f_{i+1}) \mid i \in [n]) \text{ and } ((f_j^*,q_j^*,f_{j+1}^*) \mid j \in [k])$$

are maximal chains of iterated quotients of $f: A \longrightarrow B$, then $n = k$ and there exists a sequence $(h_i: Q_i \longrightarrow Q_i^* \mid i \in [n+1])$ of isomorphisms with $h_1 = Id_A$, $q_i^* h_{i+1} = h_i \circ q_i$ for $i=2,3,\dots,n$, and $f_i^* = h_i \circ f_i$ for every $i \in [n+1]$. This can easily be seen for $i=2$ and can be proved by induction for all $i \in [n+1]$.

If $((f_i, q_i, f_{i+1}) \mid i \in \mathbb{N})$ is a maximal chain of iterated quotients of $f: A \longrightarrow B$, then any other maximal chain of iterated quotients of $f: A \longrightarrow B$ is infinite, too, and is elementwise isomorphic to the given one, in the same way as described above. Because limits of isomorphic chains are also isomorphic, we can start with any maximal chain of iterated quotients of $f: A \longrightarrow B$, which may be given by

$$((f_i, q_i, f_{i+1}) \mid i \in J) .$$

Furthermore, let L be a limit of the chain $(q_i : Q_i \longrightarrow Q_{i+1} \mid i \in J)$ and $(h_i : Q_i \longrightarrow L \mid i \in J)$ may denote the family of canonical homomorphisms. We can assume that L is constructed as the limit in Proof 3.4.4. Without loss of generality we can suppose $\langle A f \rangle = B$.
Then there exists a homomorphism

$$h: L \longrightarrow B \quad \text{with} \quad f_i = h_i \circ h \quad \text{for every } i \in J .$$

According to Theorem 2.4.2 it is sufficient to show that $h: L \longrightarrow B$ is injective and surjective.
Let $[x, i_1], [y, i_1] \in L_s$, $s \in S$, with $[x, i_1] h_s = [y, i_2] h_s$.
Because $f_i = h_i \circ h$ for every $i \in J$ and $x h_i = [x, i]$, we obtain $(x h_{i_1})h = (y h_{i_2})h$, i.e. $x f_{i_1} = y f_{i_2}$. If $i_1 < i_2$ we set

$$q: q_{i_1} \circ q_{i_1+1} \circ \dots \circ q_{i_2-1} : Q_{i_1} \longrightarrow Q_{i_2}$$

and obtain $x q f_{i_2} = x f_{i_1} = y f_{i_2}$ so that $(x, y) \in \ker f_{i_2}$.
Since q_{i_2} is natural for $\ker f_{i_2}$ it follows $x q_{i_2} = y q_{i_2}$.
Hence, by definition of L we obtain $(x, i_1) \equiv_s (y, i_2)$, i.e. $[x, i_1] = [y, i_2]$ and so $h: L \longrightarrow B$ is injective.

Now let $b \in B_s$, $s \in S$, be any element of the homomorphic image.
Because $B = \langle A f \rangle$, Corollary 2.4.7 yields the existence of

$$w: [n] \longrightarrow S, \quad t \in T(\Sigma, w)_s, \quad \underline{a} \in A_w \quad \text{with} \quad \underline{a} f_w \in \operatorname{dom} t^B$$

and $b = (\underline{a} f_w) t^B$.

Because each injective homomorphism between equoids reflects solutions of domain conditions, see Theorem 2.4.2, and

$$(\underline{a} (h_1)_w) h_w = \underline{a} f_w \in \operatorname{dom} t^B$$

it follows

$$((\underline{a}\,(h_1)_w)\,t^L)\,h = ((\underline{a}\,(h_1)_w)\,h_w)\,t^B = (\underline{a}\,f_w)\,t^B = b\,.$$

This proves that $h\colon L \longrightarrow B$ is surjective . ■

If we try to prove the theorem of homomorphisms without the hierarchy condition we fail . The bijective homomorphism

$$e\colon F^* \longrightarrow F$$

of Proof 2.4.2 can be used for showing that the theorem of homomorphisms is not true in the non-hierarchic case. However, this assertion requires the definition of the notion of homomorphic image for the non-hierarchic case. If we define the homomorphic image $\langle A\,f\rangle$ to be the subalgebra of B generated by the set-theoretic image $A\,f$, recall that then $\langle A\,f\rangle$ is closed with respect to all fundamental operations of B, then we obtain for the special homomorphism $e\colon F^* \longrightarrow F$ the equality $F^*e = F$. For being injective the identity $Id_{F^*}\colon F^* \longrightarrow F^*$ is natural for $\ker e$, so that F^* is the limit of the chain of iterated quotients of $e\colon F^* \longrightarrow F$. Since F and F^* are not isomorphic the theorem of homomorphisms is not true.

Because maximal chains of iterated quotients of a homomorphism

$$f\colon A \longrightarrow B$$

are of equal length, we define the h o m o m o r p h i c n u m b e r 'hom f' for $f\colon A \longrightarrow B$ to be the length of a maximal chain of iterated quotients. If this chain is infinite we set $\hom f = \infty$.
For each hep-signature Θ and each set of elementary implications we define $\hom(\Theta,\mathcal{A})$ to be the supremum of all $\hom f$ of homomorphisms $f\colon A \longrightarrow B$ between $\mathcal{A}$-equoids.

It seems to be of interest to give a characterization of those $(\Theta,\mathcal{A})$ that have a finite homomorphic number. However, we do not know any condition which is necessary or sufficient for the finiteness of $\hom(\Theta,\mathcal{A})$. We guess that it is recursively undecidable whether $\hom(\Theta,\mathcal{A})$ is finite or not.

3.5. Relatively free partial algebras

The aim of this section is the development of a suitable algebraic basis for parameterized abstract data types with partial operations. Examples of parameterized abstract data types are given in Chapter 1 . There the concept of an abstract data type is intuitively based on the understanding of a data type as a conceptual unit of a set, the set of admissible values, and a family of fundamental operations on the set of admissible values, where the fundamental operations or combinations of them are the only means for dealing with admissible values. Thus a data type can very well be understood as an algebraic structure. According to that understanding, parameterized data types can be represented by constructions of algebras from given ones, where the base algebras and the constructed algebras have different types in general.

In the first half of this section we introduce appropriate notions for dealing with algebras of different types.

Assume

$$\varphi = (\varphi_S, \varphi_\Omega): (S_1, \alpha_1 : \Omega_1 \longrightarrow S_1^* \times S_1) \longrightarrow$$
$$\longrightarrow (S_2, \alpha_2 : \Omega_2 \longrightarrow S_2^* \times S_2)$$

is a signature morphism (see 2.2.2), $v: X \longrightarrow S_1$ an S_1-sorted system of variables, and

$$t \in T(\Sigma_1, v)_s \quad \text{with} \quad s \in S_1 .$$

Then there is induced an S_2-sorted system of variables

$$v\varphi_S : X \longrightarrow S_2$$

resulting from the composition of the functions $v: X \longrightarrow S_1$, $\varphi_S : S_1 \longrightarrow S_2$. In the following we denote this induced S_2-sorted system of variables briefly by

$$v\varphi : X \longrightarrow S_2 .$$

Additionally, there is given an S_1-indexed family of functions (3.5.1)

$$\varphi_s : T(\Sigma_1, v)_s \longrightarrow T(\Sigma_2, v\varphi)_{s\varphi_S}, \ s \in S_1,$$

which are defined inductively as follows:

(a) If $x \in X$ then $x\varphi_{xv} = x \in T(\Sigma_2, v\varphi)_{(xv)\varphi_S}$;

(b) If $\sigma \in \Omega_1$, $\sigma: \lambda \longrightarrow s$, then
$\sigma\varphi_s = \sigma\varphi_\Omega \in T(\Sigma_2, v\varphi)_{s\varphi_S}$;

(c) If $\sigma \in \Omega_1$, $\sigma: w \longrightarrow s$, $\underline{t} \in T(\Sigma_1, v)_w$ and $t = \underline{t}\,\sigma^{T(\Sigma_1, v)}$
then
$t\varphi_s = (\underline{t}\varphi_w)(\sigma\varphi_\Omega)^{T(\Sigma_2, v\varphi)}$. ∎

In the following we allow the omission of subscripts for the functions defined in (3.5.1).

We extend these functions from terms to existence equations, sets of existence equations, and elementary implications by setting

$$(v:\ t_1 \overset{e}{=} t_2)\varphi = (v\varphi :\ t_1\varphi \overset{e}{=} t_2\varphi)\ ,$$
$$(v:\emptyset)\varphi = (v\varphi:\emptyset)\ ,$$
$$(v:G)\varphi = \{(v\varphi:\ t_1\varphi \overset{e}{=} t_2\varphi) \mid (v:\ t_1 \overset{e}{=} t_2) \in G\}\ ,$$
$$(v:\ G \longrightarrow t_1 \overset{e}{=} t_2)\varphi = (v\varphi :\ G\varphi \longrightarrow t_1\varphi \overset{e}{=} t_2\varphi)\ .$$

Next we extend the notion of a signature morphism to hep-theories. For any hep-theory T we denote by

$$S(T),\ \Omega(T),\ \Theta(T),\ \mathcal{A}(T)$$

the set of sort names of T, the set of operators of T, the hep-signature of T, and the stable set of elementary implications of T, respectively.

<u>Definition 3.5.2</u>. A t h e o r y - m o r p h i s m

$$\varphi : T_1 \longrightarrow T_2$$

is given by a signature morphism $\varphi : \Sigma(T_1) \longrightarrow \Sigma(T_2)$ so that

(1) $\mathcal{A}(T_2) \vdash (w\varphi :\ \mathrm{def}\,(\sigma\varphi) \longrightarrow (\mathrm{def}\,\sigma)\varphi)$ and
$\mathcal{A}(T_2) \vdash (w\varphi :(\mathrm{def}\,\sigma)\varphi \longrightarrow \mathrm{def}\,(\sigma\varphi))$
for every $\sigma \in \Omega(T_1)$, $\sigma : w \longrightarrow s$,

(2) $\mathcal{A}(T_2) \vdash (v:\ G \longrightarrow t_1 \overset{e}{=} t_2)\varphi$ for every
$(v:\ G \longrightarrow t_1 \overset{e}{=} t_2) \in \mathcal{A}(T_1)$. ∎

For a theory-morphism $\varphi : T_1 \longrightarrow T_2$ it is not required that $\mathrm{def}\,(\sigma\varphi)$ is equal to the image of $\mathrm{def}\,\sigma$ by φ, but it is required that $\mathrm{def}\,(\sigma\varphi)$ and $(\mathrm{def}\,\sigma)\varphi$ are semantically equivalent with respect to $\mathcal{A}(T_2)$. Analogously we do not suppose that φ maps axioms of T_1 to axioms of T_2, but we require that φ maps an axiom of T_1 to an elementary implication derivable from $\mathcal{A}(T_2)$.

Next we demonstrate how a T_1-equoid $A\varphi$ from a T_2-equoid A can be derived for a given theory-morphism $\varphi : T_1 \longrightarrow T_2$. Roughly speaking $A\varphi$ is derived by taking for every sort name $s \in S(T_1)$ the $s\varphi_S$-carrier of A and for every operator $\sigma \in \Omega(T_1)$ the fundamental operation $(\sigma\varphi_\Omega)^A$.

<u>Definition 3.5.3</u>. Let $\varphi : T_1 \longrightarrow T_2$ be a theory-morphism and A any T_2-equoid. The partial $\Sigma(T_1)$-algebra

$$A\varphi = ((A_{s\varphi} \mid s \in S(T_1)),((\sigma\varphi)^A \mid \sigma \in \Omega(T_1))$$

is said to be the φ - p a r t of the T_2-equoid A.

If $f: A \longrightarrow B$ is a homomorphism between T_2-equoids, the $S(T_1)$-indexed family of $S(T_1)$-functions

$$f\varphi = (f_{s\varphi} : A_{s\varphi} \longrightarrow B_{s\varphi} \mid s \in S(T_1))$$

is called the φ - part of the T_2-homomorphism $f: A \longrightarrow B$. ∎

Definition 3.5.3 would not fit in our approach if the φ-part of a T_2-equoid would not be a T_1-equoid. By the following corollary the preceding definition is justified.

<u>Corollary 3.5.4</u>. For every theory-morphism $\varphi: T_1 \longrightarrow T_2$ and for every T_2-equoid A the φ-part $A\varphi$ is a T_1-equoid.
For every T_2-homomorphism $f: A \longrightarrow B$ the φ-part

$$f\varphi : A\varphi \longrightarrow B\varphi$$

is a homomorphism between T_1-equoids.

<u>Proof</u>. First we point out that for every $S(T_1)$-sorted system of variables $v: X \longrightarrow S(T_1)$ it can be proved easily by induction that

$$t^{(A\varphi)} = (t\varphi)^A$$

holds for every term $t \in T(\Sigma(T_1), v)$.

An application to existence equations and sets of existence equations leads to

$$(A\varphi)_{(v:t_1 \overset{e}{=} t_2)} = A_{(v:t_1 \overset{e}{=} t_2)\varphi} \quad \text{and}$$

$$(A\varphi)_{(v:G)} = A_{(v:G)\varphi} .$$

With these identities we can prove that $A\varphi$ is a $\Theta(T_1)$-equoid.
Because of

$$(T_2) \vdash (w\varphi : (\mathrm{def}\,\sigma)\varphi \longleftrightarrow \mathrm{def}\,(\sigma\varphi))$$

for every $\sigma \in \Omega(T_1)$, $\sigma : w \longrightarrow s$, we obtain by Theorem 3.1.9 that

$$A_{(\mathrm{def}\,\sigma)\varphi} = A_{\mathrm{def}\,(\sigma\varphi)}$$

holds for every T_2-equoid, and so

$$\mathrm{dom}\,\sigma^{(A\varphi)} = \mathrm{dom}\,(\sigma\varphi)^A = A_{\mathrm{def}\,(\sigma\varphi)} = A_{(\mathrm{def}\,\sigma)\varphi} = (A\varphi)_{\mathrm{def}\,\sigma}$$

for every $\sigma \in \Omega(T_1)$, i.e. $A\varphi$ is a $\Theta(T_1)$-equoid.

Now let A by any T_2-equoid and $(v: G \longrightarrow t_1 \overset{e}{=} t_2) \in \mathcal{A}(T_1)$ any axiom of T_1. Because of condition (2) of (3.5.2)

$$\mathcal{A}(T_2) \vdash (v\varphi : G\varphi \longrightarrow t_1\varphi \overset{e}{=} t_2\varphi)$$

holds, which yields by use of Theorem 3.1.9

$$A_{(v:G)\varphi} \subseteq A_{(v:t_1 \overset{e}{=} t_2)\varphi}$$

for every T_2-equoid A. By means of the preceding identities we get

$$(A\varphi)_{(v:G)} \subseteq (A\varphi)_{(v:t_1 \overset{e}{=} t_2)}$$

and so $A\varphi$ satisfies each axiom of T_1, i.e. $A\varphi$ is a T_1-equoid.

Now we assume that $f: A \longrightarrow B$ is any T_2-homomorphism, $\sigma \in \Omega(T_1)$ with $\sigma: w \longrightarrow s$, and $\underline{a} \in \operatorname{dom} \sigma^{A^2}$.

Due to

$$\operatorname{dom} \sigma^{(A\varphi)} = \operatorname{dom} (\sigma\varphi)^A$$

we obtain

$$\underline{a}\, f_{w\varphi} \in \operatorname{dom} (\sigma\varphi)^B = \operatorname{dom} \sigma^{(B\varphi)},$$

so that

$$\underline{a}\,(f\varphi)_w = \underline{a}\, f_{w\varphi} \in \operatorname{dom} \sigma^{(B\varphi)}.$$

Finally the following holds

$$(\underline{a}\,\sigma^{(A\varphi)})(f\varphi)_s = (\underline{a}\,\sigma^{(A\varphi)})\, f_{s\varphi} = (\underline{a}(\sigma\varphi)^A)\, f_{s\varphi} =$$

$$= (\underline{a}\, f_{w\varphi})(\sigma\varphi)^B = (\underline{a}\, f_{w\varphi})\, \sigma^{(B\varphi)} = (\underline{a}(f\varphi)_w)\, \sigma^{(B\varphi)},$$

i.e. $f\varphi : A\varphi \longrightarrow B\varphi$ is a T_1-homomorphism. ∎

According to Definition 3.5.2 a hep-theory T_1 is said to be a s u b t h e o r y of a hep-theory T_2 if $S(T_1) \subseteq S(T_2)$, $\Omega(T_1) \subseteq \Omega(T_2)$ and if the inclusions form a theory-morphism, i.e.

$$\mathcal{A}(T_2) \vdash (w: \operatorname{def}_{T_1} \sigma \longleftrightarrow \operatorname{def}_{T_2} \sigma) \quad \text{for every } \sigma \in \Omega(T_1),$$

and

$$\mathcal{A}(T_2) \vdash (v: G \longrightarrow t_1 \overset{e}{=} t_2)\varphi$$

for every $(v: G \longrightarrow t_1 \overset{e}{=} t_2) \in \mathcal{A}(T_1)$.

Also in the case of a subtheory we do not require the inclusion $\mathcal{A}(T_1) \subseteq \mathcal{A}(T_2)$ and the identity $\operatorname{def}_{T_1} \sigma = \operatorname{def}_{T_2} \sigma$ although these will be true usually.

If T_1 is a subtheory of T_2 this will be denoted briefly by

$$T_1 \subseteq T_2.$$

For the sake of simplicity we do not use explicitly an inclusion morphism in the case of a subtheory, but we use the following notion:

$$A{\downarrow}T_1 \quad \text{and}$$

$$f{\downarrow}T_1 : A{\downarrow}T_1 \longrightarrow B{\downarrow}T_1$$

for every T_2-equoid A and every T_2-homomorphism $f: A \longrightarrow B$, respectively. We speak of the T_1-part $A{\downarrow}T_1$ of the T_2-equoid A or of the underlying T_1-equoid $A{\downarrow}T_1$ of the T_2-equoid A.

For instance, the theory of directed graphs

```
T_GRAPH is  sorts  Edges, Nodes
        oprn  source(Edges) ---> Nodes
              target(Edges) ---> Nodes
        axioms  none
   end  T_GRAPH
```

corresponds to a subtheory of the theory T_{CAT} of small categories defined in Section 2.5. This correspondence is given by the theory-morphism

$$\varphi : T_{GRAPH} \longrightarrow T_{CAT}$$

given by

$$\text{Edges}\,\varphi = \text{Mor},\quad \text{Nodes}\,\varphi = \text{Obj},$$
$$\text{source}\,\varphi = \text{source},\quad \text{target}\,\varphi = \text{target}\ .$$

Thus, for every small category $C \in ALG(T_{CAT})$ the φ-part $C\varphi$ is a directed graph, where the objects of C are interpreted as the nodes of $C\varphi$ and the morphisms of C are interpreted as the edges of $C\varphi$.

The construction of the φ-part $A\varphi$ for a theory-morphism $\varphi: T_1 \longrightarrow T_2$ and a T_2-equoid A suggests looking for ways of inverting this construction, i.e. looking for ways to construct a T_2-equoid for every given T_1-equoid and every theory-morphism $\varphi : T_1 \longrightarrow T_2$ in a canonical way.

One way to extend a T_1-equoid B to a T_2-equoid with respect to a theory-morphism $\varphi : T_1 \longrightarrow T_2$ is given by the following definition.

<u>Definition 3.5.5</u> Let $\varphi: T_1 \longrightarrow T_2$ be a theory-morphism. A T_2-equoid F is said to be f r e e l y g e n e r a t e d by a T_1-equoid B according $\varphi: T_1 \longrightarrow T_2$ if

(1) there is a T_1-homomorphism $h: B \longrightarrow F$, such that

(2) for every T_2-equoid A and every T_1-homomorphism $f: B \longrightarrow A$ there is exactly one T_2-homomorphism $f^{\#}: F \longrightarrow A$ with $f = h\circ(f^{\#}\varphi)$. ∎

To illustrate this concept we discuss some special cases. If T_1

is the e m p t y t h e o r y denoted by $T_\emptyset$, then this theory has exactly one model, i.e. there is exactly one $T_\emptyset$-equoid namely the empty sequence of carrier sets together with the empty sequence of fundamental operations. This uniquely determined $T_\emptyset$-equoid will also be denoted by $\emptyset$ so that $ALG(T_\emptyset) = \{\emptyset\}$. A T-equoid F freely generated by $\emptyset$ according to the inclusion $T_\emptyset \subseteq T$ is then characterized by the fact that for every T-equoid A exactly one T-homomorphism $f: F \longrightarrow A$ exists. We see that in this special case the notion introduced coincides with the concept of an initial T-equoid.

Now let us assume that $T_o \subseteq T$ with $S(T_o) = S(T) = S$ and $\Omega(T_o) = \emptyset$. In this case T_o-equoids are exactly S-sets and T-homomorphisms are exactly S-functions. A T-equoid freely generated by an S-set $X = (X_s \mid s \in S)$ according to the inclusion $T_o \subseteq T$ is then characterized by the existence of an S-function $h: X \longrightarrow F{\downarrow}T_o$ such that for every T-equoid A and every S-function $f: X \longrightarrow A{\downarrow}T_o$ exactly one T-homomorphism $f^{\#}: F \longrightarrow A$ exists with $f = h \circ (f^{\#}{\downarrow}T_o)$. Since the construction $A{\downarrow}T_o$ associates with every T-equoid A its S-carrier and $f^{\#}{\downarrow}T_o = f^{\#}$ the notion introduced coincides with the well-known concept of a T-equoid freely generated by an S-set X.

More general, the notion introduced allows us to speak of an Abelian group F freely generated by a group G which is implemented by $F = G/K$ where K is the commutator of G.

With respect to partial algebras we can also speak of a small category freely generated by a directed graph. This construction can be implemented by the so-called category of finite paths over a directed graph (see Hasse,Michler (1966)).

From Definition 3.5.5 it follows immediately that a T_2-equoid F freely generated by a T_1-equoid B according $\varphi: T_1 \longrightarrow T_2$ is determined uniquely up to isomorphism.

By means of the calculus introduced in Section 3.2 we can prove that for every theory-morphism $\varphi: T_1 \longrightarrow T_2$ relatively free equoids exist or in terms of the theory of categories that the forgetful functor $A \longmapsto A\varphi$, for every T_2-equoid A, has a left-adjoint functor.

<u>Theorem 3.5.6</u>. For every theory-morphism $\varphi: T_1 \longrightarrow T_2$ and every T_1-equoid B there exists a T_2-equoid F freely generated by B according $\varphi: T_1 \longrightarrow T_2$.

Proof. In Section 3.3 we proved that any T_1-equoid B is isomorphic to a T_1-equoid $F(\mathcal{O}(T_1),G,v)$ with an appropriate set of existence equations $(v:G)$, $v\colon X \longrightarrow S(T_1)$. Consequently, we can suppose

$$B = F(\mathcal{O}(T_1),G,v)$$

without loss of generality. Taking

$$F = F(\mathcal{O}(T_2),G\varphi,v\varphi) \in ALG(T_2) \quad \text{and} \quad \underline{e} \in F_{(v:G)}$$

as universal solution we obtain

$$\underline{e} \in F_{(v:G)\varphi} = (F\varphi)_{(v:G)},$$

due to Proof 3.5.4.
If $\underline{1} \in F(\mathcal{O}(T_1),G,v)_{(v:G)}$ denotes the universal solution it follows that there is exactly one homomorphism $h\colon B \longrightarrow F\varphi$ with $\underline{1}\,h_v = \underline{e}$.
Now let be given any $A \in ALG(T_2)$ and $f\colon B \longrightarrow A\varphi$. Since a homomorphism preserves solutions of $(v:G)$ we may infer

$$\underline{1}\,f_v \in (A\varphi)_{(v:G)} = A_{(v:G)\varphi},$$

so that there is exactly one T_2-homomorphism

$$f^{\#}\colon F \longrightarrow A \quad \text{with} \quad \underline{e}\,f^{\#}_{v\varphi} = \underline{1}\,f_v.$$

This identity leads to

$$\underline{1}(h\circ(f^{\#}\varphi))_v = \underline{1}(h_v\circ(f^{\#}\varphi)_v) = \underline{1}(h_v\circ f^{\#}_{v\varphi}) = (\underline{1}\,h_v)(f^{\#}_{v\varphi}) = \underline{e}\,f^{\#}_{v\varphi} =$$
$$= \underline{1}\,f_v$$

and so $h\circ(f^{\#}\varphi) = f$.
If $\hat{f}\colon F \longrightarrow A$ is any T_2-homomorphism with $h\circ(\hat{f}\varphi) = f$ we can conclude

$$\underline{e}\,f^{\#}_v = \underline{1}\,f_v = \underline{1}(h\circ(\hat{f}\varphi))_v = (\underline{1}\,h_v)(\hat{f}\varphi)_v = \underline{e}(\hat{f}\varphi)_v = \underline{e}\,\hat{f}_{v\varphi}$$

which involves

$$f = f^{\#}. \qquad \blacksquare$$

We mentioned previously that the construction of the category of paths over a directed graph forms an example for (3.5.6). To make this visible we start with a sorted system of variables

$$w\colon \{x_1,x_2,x_3,x_4,x_5\} \longrightarrow \{\text{Edges},\text{Nodes}\}$$

with $x_i w = \text{Edges}$ for $i=1,2,3,4,5$
and with a set of existence equations

$$(w:G) = (w:\{\ \text{source}(x_1) \overset{e}{=} \text{source}(x_2),\ \text{target}(x_2) \overset{e}{=} \text{source}(x_3)$$
$$\text{target}(x_3) \overset{e}{=} \text{source}(x_5),\ \text{source}(x_3) \overset{e}{=} \text{source}(x_4)$$
$$\text{target}(x_5) = \text{target}(x_4)\}\ ).$$

A directed graph G freely generated by (w:G) is graphically described as follows:

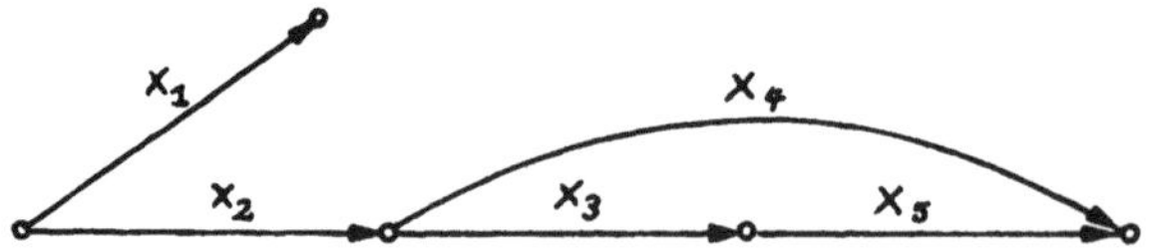

Taking the theory-morphism

$\varphi : T_{GRAPH} \longrightarrow T_{CAT}$

as defined above it results in

$w\varphi : \{x_1,\ldots,x_5\} \longrightarrow \{Obj, Mor\}$

with $x_i w\varphi$ = Mor for i=1,2,...,5 and (w:G)

so that the small category C freely generated by the directed graph G according $\varphi : T_{GRAPH} \longrightarrow T_{CAT}$ is given by

$C = F(\mathfrak{A}_{CAT}, G\varphi, w\varphi)$

which can be represented by the following diagram:

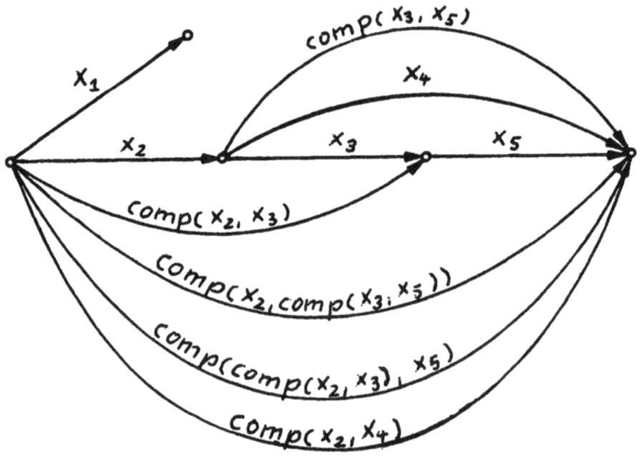

Every additional morphism corresponds to a non-trivial, finite path. In our example there are four such paths.

We conclude this section with a second example. For this reason let be given the theory T_2 with a subtheory T_1

T_1 <u>is</u> <u>sorts</u> Bool, Elements
<u>oprn</u> t,f $\longrightarrow$ Bool
equal(Elements,Elements) $\longrightarrow$ Bool
<u>axioms</u> x,y: Elements

$equal(x,x) = t$
if $equal(x,y) = t$ **then** $x = y$
end T_1

T_2 **is** T_1 **with**
sorts Sets
oprn $1 \longrightarrow$ Sets
$ap(Sets, Elements) \longrightarrow Sets$
axioms x,y: Elements, m: Sets
$ap(ap(m,x),x) = ap(m,x)$
$ap(ap(m,x),y) = ap(ap(m,y),x)$
end T_2

This notation of hep-theories, similar to the notation of Chapter 1, differs from the formal definition which requires that every axiom is of the form

$$(v\colon G \longrightarrow t_1 \overset{e}{=} t_2)\ .$$

First, we can in almost all examples assume that all axioms have the same S-sorted system of variables. With the declaration

x,y: Elements, m: Sets

we define for instance the following S-sorted system

$$w\colon \{x,y,m\} \longrightarrow \{Elements, Sets, Bool\}$$

with $xw = yw =$ Elements and $mw =$ Sets. However, we havenot introduced the identifier 'w'.
Secondly, if in $(v\colon G \longrightarrow t_1 \overset{e}{=} t_2)$ the premise G is empty, we write briefly $t_1 = t_2$ and if G is not empty then we use the notation

if G **then** $t_1 = t_2$.

Now let us assume that $B \in ALG(T_1)$ with

$B_{Bool} = \{T,F\}$, $B_{Elements} = M$ any set ,
$t^B = T$, $f^B = F$, and
$equal^B(x,y) = T$ iff $x = y$ for all $x,y \in M$.

Of course, not every T_1-equoid is of this kind. But,if we choose the T_1-equoid in this kind, then there is a nice description of a T_2-equoid F freely generated by B according to the inclusion $T_1 \subseteq T_2$. We can take

$F_{Bool} = B_{Bool}$, $F_{Elements} = B_{Elements}$, $t^F = t^B$, $f^F = f^B$, and
$equal^F = equal^B$, i.e. $F{\downarrow}T_1 = B$.

The remaining sort name and the remaining operators of T_2 will be interpreted as follows:

$$F_{Sets} = \text{set of all finite subsets of } M = F_{Elements} ,$$
$$1^F = \emptyset \in F_{Sets} ,$$
$$ap^F(m,x) = m \cup \{x\} \quad \text{for every } m \in F_{Sets},\ x \in F_{Elements} .$$

By induction on the number of elements of finite subsets one can prove that F is freely generated by B.

Characteristic for this example is that T_2 has a new sort name and no new operators whose range is a sort name of T_1. All new operators generate values of an additional sort. This property of the theory extension $T_1 \subseteq T_2$ guarantees that for every T_2-equoid F freely generated by a T_1-equoid B the identity $B = F{\downarrow}T_1$ holds. This property of a theory extension will be of great importance for the next chapter, i.e. for the applicability of Theorem 3.5.6 to parameterized abstract data types as described in Chapter 1.

Now we prove a property of relatively free equoids that enables us to build specifications of abstract data types step by step.

Corollary 3.5.7. Let $\varphi: T_1 \longrightarrow T_2$, $\psi: T_2 \longrightarrow T_3$ be theorymorphisms. If $F \in ALG(T_2)$ is freely generated by $B \in ALG(T_1)$ according $\varphi: T_1 \longrightarrow T_2$, and if $F^{\#} \in ALG(T_3)$ is freely generated by F according $\psi: T_2 \longrightarrow T_3$, then $F^{\#}$ is freely generated by B according $\varphi\circ\psi: T_1 \longrightarrow T_3$.

Proof. Let $h: B \longrightarrow F\varphi$, $g: F \longrightarrow F^{\#}\psi$ be the homomorphisms corresponding to Definition 3.5.5. Then

$$h\circ(g\varphi): B \longrightarrow (F^{\#}\psi)\varphi = F^{\#}(\varphi\circ\psi) .$$

If $A \in ALG(T_3)$ and $f: B \longrightarrow A(\varphi\circ\psi) = (A\psi)\varphi$, then there is exactly one homomorphism

$$f^{\#}: F \longrightarrow A\psi \quad \text{with} \quad h\circ(f^{\#}\varphi) = f .$$

This involves the existence of exactly one homomorphism

$$f^{\#\#}: F^{\#} \longrightarrow A \quad \text{with} \quad g\circ(f^{\#\#}\psi) = f^{\#}$$

so that

$$f^{\#}\varphi = (g\circ(f^{\#\#}\psi))\varphi = (g\varphi)\circ(f^{\#\#}\psi)\varphi$$

and therefore

$$(h\circ(g\varphi))(f^{\#\#}(\varphi\circ\psi)) = h\circ(g\varphi)((f^{\#\#}\psi)\varphi) = h\circ(f^{\#}\varphi) = f . \qquad \blacksquare$$

We illustrate the application of Corollary 3.5.7 by the following extension $T_2 \subseteq T_3$:

```
T3  is  T2  with
        oprn  el(Elements,Sets) ——> Bool
        axioms  x,y: Elements, m: Sets
        el(x,l) = f
        if el(x,m) = f then  el(y,ap(m,x)) = equal(x,y)
        if el(x,m) = t then  el(x,ap(m,y)) = t
    end  T3
```

Let $B \in ALG(T_1)$, $F \in ALG(T_2)$ be as above. A T_3-equoid $F^{\#}$ freely generated by F according $T_2 \subseteq T_3$ is characterized by

$$F^{\#}\downarrow T_2 = F \quad \text{and} \quad el^{F^{\#}} = \text{element relation.}$$

By Corollary 3.5.7 $F^{\#}$ is also freely generated by B according $T_1 \subseteq T_3$.

CHAPTER 4

CANONS - INITIALLY RESTRICTING ALGEBRAIC THEORIES

The algebraic approach developed in Chapter 2 and Chapter 3 is not sufficient for descriptions of those classes of partial algebras that appear as model classes of specifications in Chapter 1. Especially the initial satisfaction of axioms as used for enrichments by definition cannot be expressed by the algebraic notions developed in the preceding two chapters. It is the aim of this chapter to introduce suitable notions, so-called initial restrictions in algebraic theories.

In the first section we define the concept of a CANON and investigate classes of models of canons. In a slightly restricted form the notion of a canon was introduced in Kaphengst, Reichel (1971). Over an intermediate form in Reichel (1979) the final form was defined in Reichel (1980) following a suggestion of Kaphengst. Quite the same concept was used in Burstall, Goguen (1980) for the definition of the semantics of the specification language CLEAR. Later Ehrig, see Ehrig (1981), used an extension of this concept for dealing with parameterized data types. But it is worth to mention here that the full expressive power of the canon concept can only be achieved by the use of partial algebras. This will be made comprehensible in Section 4.2 which is aimed to the development of a formalization of the concept of an abstract algorithm over an parameterized abstract data type. Beside a justification of this notion we will prove a necessary condition for a function over a parameterized abstract data type to be an algorithm, i.e. a computable parameterized function. Additionally we investigate uniquely specifiable functions over parameterized abstract data types. The difference between this both concepts is demonstrated in Chapter 1 by the binary relation in the set of nodes of a directed graph indicating whether two nodes are connected by a finite directed path. This relation is an example of a uniquely specifiable but not computable function.

In the last section of this chapter we introduce morphisms between canons and prove some of their properties. This finishes the set of formal notions necessary for the semantics of the specifications given in Chapter 1. Since the introduction of a specification language is not the aim of this book we will not give a complete and formal definition of the semantics. But there are no serious problems for this task since we made available all tools for inductive definitions even if the production rules, i.e. the generating constructions, are restricted by context conditions.

4.1 Canon Model Classes

The essential difference between model classes of $(\Theta,\mathcal{A})$-algebras, where Θ is an ep-signature and $\mathcal{A}$ a stable set of elementary implications, and model classes of specifications containing enrichments by definition can be illustrated by the example of finite directed graphs where each edge is valuated by a natural number. An axiomatization of this class of structures requires three sorts of objects, namely nodes, edges and natural numbers. Beside the both unary operations 'begin' and 'end' a third operation 'valuation' associates with every edge a natural number. This class of valuated finite directed graphs cannot be described by a first order theory. This impossibility is an immediate consequence of the cardinality theorem ,see Shoenfield (1967). On the other hand this class is a typical class of models of a specification as the following specification shows.

V-D-GRAPH is D-GRAPH with NAT
with definition
sorts E-Sets, N-Sets
oprn e-empty ⟶ E-Sets
n-empty ⟶ N-Sets
e-join(E-Sets,Edges) ⟶ E-Sets
n-join(N-Sets,Nodes) ⟶ N-Sets
axioms x_1,x_2: Edges, y_1,y_2: Nodes, m: E-Sets, k: N-Sets
$e\text{-}join(e\text{-}join(m,x_1),x_1) = e\text{-}join(m,x_1)$
$n\text{-}join(n\text{-}join(k,y_1),y_1) = n\text{-}join(k,y_1)$
$e\text{-}join(e\text{-}join(m,x_1),x_2) = e\text{-}join(e\text{-}join(m,x_2),x_1)$
$n\text{-}join(n\text{-}join(k,y_1),y_2) = n\text{-}join(n\text{-}join(k,y_2),y_1)$
with requirement
oprn edges ⟶ E-Sets
nodes ⟶ N-Sets
axioms x: Edges, y: Nodes
e-join(edges,x) = edges
n-join(nodes,y) = nodes
with requirement

<u>oprn</u> valuation(Edges) $\longrightarrow$ Nat

<u>end</u> V-D-GRAPH

The preceding manner of specifying finite structures was suggested by Kaphengst in a personal communication.

Theorem 3.5.6 is the key for the following notion of an initial restriction in an algebraic theory.

<u>Definition 4.1.1</u>. Let be given algebraic theories T, T_2, a subtheory T_1 of T_2, and a theory morphism $\varphi: T_2 \longrightarrow T$, see 3.5.2. The ordered pair

$$(T_1, \varphi: T_2 \longrightarrow T)$$

is called an i n i t i a l r e s t r i c t i o n in T.
A T-algebra A satisfies this initial restriction, denoted by

$$A \models (T_1, \varphi: T_2 \longrightarrow T)$$

if the φ-part $A\varphi$ is freely generated by $(A\varphi){\downarrow}T_1$ according to the inclusion $T_1 \subseteq T_2$, see 3.5.5. ∎

We draw attention to the following fact. If

$$A \models (T_1, \varphi: T_2 \longrightarrow T)$$

then $A\varphi$ is not only freely generated by $(A\varphi){\downarrow}T_1$ in the sense of (3.5.5) but $A\varphi$ is a free T_2- e x t e n s i o n of $(A\varphi){\downarrow}T_1$ with respect to the inclusion $T_1 \subseteq T_2$. With respect to an inclusion $T_1 \subseteq T_2$ we say that a T_2-algebra F is a free T_2-extension of a T_1-algebra B if

(1) $F{\downarrow}T_1 = B$ and

(2) for every $A \in ALG(T_2)$ and every homomorphism $f: B \longrightarrow A{\downarrow}T_1$ there exists exactly one homomorphism $f^{\#}: F \longrightarrow A$ with $f = f^{\#}{\downarrow}T_1$.

For the notion of a free extension the first condition is of special interest. In general condition (1) is not satisfied, especially for one-sorted algebras. Unfortunately it is recursively undecidable if for any inclusion $T_1 \subseteq T_2$ for every T_1-algebra B a free T_2-extension exists, whereas Theorem 3.5.6 guarantees that for every T_1-algebra B there is a T_2-algebra freely generated by B according to the inclusion $T_1 \subseteq T_2$.

Let us consider two extremal cases. The first is given by the initial restriction (T,id: T $\longrightarrow$ T). This initial restriction in T is trivial since every T-algebra satisfies it. On the other hand, the initial restriction

$$(\emptyset,\ i: \emptyset \longrightarrow T)$$

where $\emptyset$ denotes the empty subtheory and $i: \emptyset \longrightarrow T$ its inclusion, is called the *total* initial restriction of T. It is easy to see that exactly the initial T-algebras satisfy the total initial restriction.

An initial restriction $(T_1, \varphi: T_2 \longrightarrow T)$ is said to be *inner* if T_2 is a subtheory of T and $\varphi: T_2 \longrightarrow T$ is the inclusion. Almost all realistic examples of initial restrictions, especially all those used in this book, are inner ones. But from a more theoretical point of view it is not possible to restrict the approach to inner initial restrictions.

Since a specification may contain several enrichments by definition we need the following notion.

Definition 4.1.2. An ordered pair

$$(T, \Delta)$$

consisting of a finitary algebraic theory T and a finite set Δ of initial restrictions in T is called a *canon*.
A T-algebra A is said to be a (T,Δ)-algebra if A satisfies every initial restriction of Δ. ALG(T,Δ) denotes the class of all (T,Δ)-algebras. ∎

Immediately we obtain

$$\mathrm{ALG}(T,\emptyset) = \mathrm{ALG}(T) \quad \text{and}$$
$$\mathrm{ALG}(T,\{(\emptyset, i:\emptyset \longrightarrow T)\}) = \text{isomorphism class of the initial T-algebra.}$$

A canon (T,Δ) is called *totally restricted* if

$$\mathrm{ALG}(T,\Delta) = \mathrm{ALG}(T,\{(\emptyset, i:\emptyset \longrightarrow T)\}) .$$

In view of this fact we can use totally restricted canons as mathematical representations of abstract data types.

As an illustration of the notion introduced we give the canon described by the specification V-D-GRAPH. For that reason we consider the following theories:

$T_{D\text{-}GRAPH}$ __is__ __sorts__ Edges, Nodes
    __oprn__ begin(Edges) ⟶ Nodes
        end(Edges) ⟶ Nodes
    __axioms__ none

T_{NAT} __is__ __sorts__ Nat, Bool
    __oprn__ true, false ⟶ Bool
        zero ⟶ Nat
        succ(Nat) ⟶ Nat
        eq(Nat,Nat) ⟶ Bool
    __axioms__ x,y: Nat
    eq(x,x) = true
    eq(x,y) = eq(y,x)
    eq(zero,succ(x)) = false
    eq(succ(x),succ(y)) = eq(x,y)

T_2 __is__ $T_{D\text{-}GRAPH}$ __with__ T_{NAT}

T_3 __is__ T_2 __with__
    __sorts__ E-Sets, N-Sets
    __oprn__ e-empty ⟶ E-Sets
        n-empty ⟶ N-Sets
        e-join(E-Sets,Edges) ⟶ E-Sets
        n-join(N-Sets,Nodes) ⟶ N-Sets
    __axioms__ x_1, x_2: Edges, y_1, y_2: Nodes, m: E-Sets, k: N-Sets
    e-join(e-join(m,x_1),x_1) = e-join(m,x_1)
    n-join(n-join(k,y_1),y_1) = n-join(k,y_1)
    e-join(e-join(m,x_1),x_2) = e-join(e-join(m,x_2),x_1)
    n-join(n-join(k,y_1),y_2) = n-join(n-join(k,y_2),y_1)

$T_{V\text{-}D\text{-}GRAPH}$ __is__ T_3 __with__
    __sorts__ no new sorts
    __oprn__ edges ⟶ E-Sets
        nodes ⟶ N-Sets
        valuation(Edges) ⟶ Nat
    __axioms__ x: Edges, y: Nodes
    e-join(edges,x) = edges
    n-join(nodes,y) = nodes

The specification V-D-GRAPH describes a canon $\mathfrak{C}_{V\text{-}D\text{-}GRAPH}$ with two initial restrictions in $T_{V\text{-}D\text{-}GRAPH}$:

$$\mathbb{C}_{V\text{-}D\text{-}GRAPH} = (T_{V\text{-}D\text{-}GRAPH}, \{(\emptyset, i_1 : T_{NAT} \longrightarrow T_{V\text{-}D\text{-}GRAPH}), (T_2, i_2 : T_3 \longrightarrow T_{V\text{-}D\text{-}GRAPH})\})$$

where i_1, i_2 denote the inclusions.

The first initial restriction defines the abstract data type of natural numbers and the abstract data type of truth values together with the equality of natural numbers. The second initial restriction defines the parameterized abstract data types of the sets of finite subsets of the sets of edges and nodes, respectively. The second initial restriction represents a parameterized abstract data type because the subtheory T_2 is not totally restricted by the first initial restriction.

The study of properties of model classes of canons is not only a natural algebraic question but also of some interest from point of view of applications. Based on those properties one can derive tests for deciding if certain constructions of sets of of functions can be specified by the initial canon specification method. By virtue of the following result we can state for instance that the general powerset construction cannot be specified and that the class of all complete lattices is not the model class of any canon.

In the following we present some results obtained by K.Benecke in his thesis for special kinds of canons and extended to general canons in Benecke (1983). We conclude this section with a new result concerning the closure under quotient constructions.

Because a model class of a canon can be the isomorphism class of an initial T-algebra model classes of canons are evidently not necessarily closed with respect to direct products.

By the following example we demonstrate that a class of models of a canon is in general not closed with respect to the intersection of subalgebras.

EXAMPLE 4.0 <u>is</u> BOOL <u>with</u> <u>requirement</u>

<u>sorts</u> M

<u>oprn</u> eq(M,M) ⟶ Bool

<u>axioms</u> x,y: M

eq(x,x) = true

<u>if</u> eq(x,y) = true <u>then</u> x = y

<u>end</u> EXAMPLE 4.0

EXAMPLE 4.1 <u>is</u> EXAMPLE 4.0 <u>with definition</u>

<u>sorts</u> Q

<u>oprn</u> q (M,M) ⟶ Q

<u>axioms</u> x,y,z: M

<u>if</u> eq(x,y)=false, eq(x,z)=false, eq(y,z)=false

<u>then</u> q(x,y)=q(x,x), q(x,x)=q(z,z) <u>fi</u>

<u>end</u> EXAMPLE 4.1

In this specification we use a non-elementary implication as equivalent of two elementary implications with identical premises. This specification describes a canon with two inner initial restrictions. The first initial restriction is given by the subspecification BOOL and the second initial restriction is given by the last enrichment by definition. This enrichment by definition implies that for every algebra

$$A \in ALG(\mathfrak{C}_{EXAMPLE\ 4.1})$$

A_Q is a duplicate of $A_M \times A_M$ provided A_M has no more than two elements. If A_M has three or more elements the implication involves that A_Q consists of exactly one element. Consequently, each algebra A of $ALG(\mathfrak{C}_{EXAMPLE\ 4.1})$ is up to isomorphism uniquely determined by the set A_M.

Let be given algebras $A,B,C \in ALG(\mathfrak{C}_{EXAMPLE\ 4.1})$ with

$$A_M = \{1,2,3,4\}, \quad A_Q = \{1\},$$
$$B_M = \{1,2,3\},$$
$$C_M = \{1,2,4\},$$
$$B_Q = C_Q = A_Q .$$

Then B and C are both subalgebras of A but the subalgebra

$$D = B \cap C$$

of A does not satisfy the second initial restriction, because $D_M = \{1,2\}$ and $D_Q = \{1\}$. D would only be a model of the canon $\mathfrak{C}_{EXAMPLE\ 4.1}$ if

$$D_Q = \{(1,1),(2,2),(1,2),(2,1)\} .$$

We see that model classes of canons have a different behaviour from varieties and quasi-varieties. Roughly speaking one can say they behave diametrically. The preceding example V-D-GRAPH demonstrates that they behave also in a different way from model classes of first order theories. Contrary to the cardinality theorem any model of V-D-GRAPH is countable and no model of greater cardinality exists.

Because of the following theorem every model class of a canon has a model of at most countable cardinality if at least one model exists. So we can say that model classes of canons satisfy a downward cardinality theorem.

<u>Theorem 4.1.3</u>. Let be given any canon (T,Δ), any (T,Δ)-algebra A, and any S-subset $M = (M_s \mid s \in S)$ of the S-carrier of A, where S denotes the set of sort names of T. Then there exists a (T,Δ)-algebra B which is a subalgebra of A so that the cardinality of B_s is the maximum of the cardinality of M_s and of the set of natural numbers for every $s \in S$.

<u>Proof</u>. If $\Delta = \emptyset$, i.e. if there are no initial restrictions, then we can construct B as the T-subequoid of A generated by M.

For the sake of simplicity we suppose that Δ consists of only one initial restriction $(T_1, \varphi: T_2 \longrightarrow T)$. In the general case one can use the same idea of the following proof, but this idea would then be hidden by notational problems.

Now let be $A \in ALG(T,\Delta)$ and $M \subseteq A$. Then we abbreviate

$$A'' = A\varphi \in ALG(T_2),$$

$$A' = A''{\downarrow}T_1 = (A\varphi){\downarrow}T_1 \in ALG(T_1)$$

so that A'' is freely generated by A'.
Without loss of generality we may suppose that the carriers of A' are pairwise disjoint. By

$$\hat{w}: \cup\{A'_s \mid s \in S(T_1)\} \longrightarrow S(T_1) \quad \text{and}$$

$$w: \cup\{A'_s \mid s \in S(T_2)\} \longrightarrow S(T_2) \supseteq S(T_1)$$

we denote the sorted systems of variables corresponding to A' taken for an $S(T_1)$-set and $S(T_2)$-set, respectively. Further let $\underline{e} \in A''_w$ and $\underline{e} \in A'_{\hat{w}}$ denote the identical inclusion of the $S(T_2)$-set A' into A'' and the identity of the $S(T_1)$-set A', respectively. Corresponding to Section 3.3 we set

$$(w{:}G_{A'}) = \{(\hat{w}: a \overset{s}{=} t) \mid t \in T(\Sigma(T_1),\hat{w})_s,\ s \in S(T_1),\ a = \underline{e}\, t^{A'}\},$$

$$(w{:}G_{A'}) = \{(w: a \overset{s}{=} t) \mid t \in T(\Sigma(T_1),w)_s,\ s \in S(T_1),\ a = \underline{e}\, t^{A'}\}.$$

Because of $A \in ALG(T,\Delta)$ we obtain by Theorem 3.5.6 that

A' is isomorphic to $F(\mathfrak{A}(T_1), G_{A'}, w)$ and
A'' is isomorphic to $F(\mathfrak{A}(T_2), G_{A'}, w)$

and that $\underline{e} \in A''_w$ is the universal solution of $(w{:}G_{A'})$ in A''.

For every $t \in T(\Sigma(T_2),w)$ there exists a smallest $S(T_2)$-sorted subsystem

$$w_t : var(t) \longrightarrow S(T_2)$$

of w with $t \in T(\Sigma(T_2),w_t)$, where $var(t)$ is the finite $S(T_2)$-set of all variables used syntactically in the term t . Analogously, for every finite set $(w:G)$ of existence equations on w there exists a smallest finite $S(T_2)$-sorted subsystem

$$w_G : var(w:G) \longrightarrow S(T_2)$$

of w with

$$t_1, t_2 \in T(\Sigma(T_2),w_G)$$

for every $(w: t_1 \overset{e}{=} t_2) \in (w:G)$.

According to Definition 3.1.7 for every $(w: t_1 \overset{e}{=} t_2)$ with

$$\mathcal{A}(T_2) \models (w: G_{A^{\cdot}} \longrightarrow t_1 \overset{e}{=} t_2)$$

there exists a finite subset of $G_{A^{\cdot}}$, denoted by $G_{(w:t_1 \overset{e}{=} t_2)}$ with

$$\mathcal{A}(T_2) \models (w: G_{(w:t_1 \overset{e}{=} t_2)} \longrightarrow t_1 \overset{e}{=} t_2) .$$

After these notational conventions we can start with the construction of the required (T,Δ)-algebra B . B will be constructed by iteration of three different closure operations starting with M .
For each natural number $k \in \mathbb{N}$ we set

(o) $M^0 = M$

(i) $M^{3k+1} = M^{3k} \cup \{ var(t) \mid t \in T(\Sigma(T_2),w)_s , s \in S(T_2) , s \notin S(T_1) , \underline{e}\, t^{A^{\cdot}} \in M_s^{3k} \}$,

(ii) $M^{3k+2} = M^{3k+1} \cup \{ var(G_{(w:t_1 \overset{e}{=} t_2)}) \mid t_1, t_2 \in T(\Sigma(T_2),w) , \underline{e} \in A^{\cdot}_{(w:t_1 \overset{e}{=} t_2)} , var(t_1) \subseteq M^{3k+1}, var(t_2) \subseteq M^{3k+1} \}$,

(iii) $M^{3k+3} = M^{3k+2} \cup \{ \underline{m}\,\sigma^A \mid \sigma \in \Omega(T),\ \sigma : v \longrightarrow s ,\ \underline{m} \in M_v^{3k+2} \}$

and finally

$$B = \bigcup_{k \in \mathbb{N}} A^{3k+3} .$$

Note that these constructions are performed with $S(T)$-sets.
For the general proof, where Δ consists of more than one initial restriction in T , we would have to modify the closure operations (i) and (ii). For every initial restriction of Δ there would be an expression analogous to the one which we have constructed.

Due to the closure operation (iii) the S(T)-subset B is closed with respect to all fundamental operations σ^A, $\sigma \in \Omega(T)$, and so B is the carrier of a T-subequoid of A.

It remains to show that B satisfies the initial restriction

$$(T_1, \varphi : T_2 \longrightarrow T).$$

Just as above we abbreviate

$$B'' = B\varphi \in ALG(T_2),$$
$$B' = (B\varphi)\downarrow T_1 \in ALG(T_1)$$

and denote by

$$\hat{u}: \cup\{B'_s \mid s \in S(T_1)\} \longrightarrow S(T_1)$$
$$u: \cup\{B'_s \mid s \in S(T_1)\} \longrightarrow S(T_2) \supseteq S(T_1)$$

the sorted systems of variables corresponding to B' taken for an $S(T_1)$-set and $S(T_2)$-set, respectively. The construction of B implies that $\hat{u}$ is a subsystem of $\hat{w}$ and u is a subsystem of w. Let $\underline{i} \in B''_u$ and $\underline{i} \in B'_{\hat{u}}$ denote the identical inclusion of the $S(T_1)$-set B' into B" and the identity of the $S(T_1)$-set B', respectively. Finally we set

$$(u:G_{B'}) = \{(u:b \overset{e}{=} t) \mid t \in T(\Sigma(T_1),\hat{u})_s,\ s \in S(T_1),\ b = \underline{i}\, t^{B'}\}$$

$$(u:G_{B'}) = \{(u:b \overset{e}{=} t) \mid t \in T(\Sigma(T_1),u)_s,\ s \in S(T_1),\ b = \underline{i}\, t^{B'}\}.$$

Contrary to the first part of this proof we have now to show that B" is isomorphic to $F(\mathcal{C}(T_2),G_{B'},u)$.

According to Theorem 3.2.4 for the solution

$$\underline{i} \in B''_{(u:G_{B'})}$$

there exists exactly one homomorphism

$$i^{\#} : F(\mathcal{C}(T_2),G_{B'},u) \longrightarrow B''$$

defined by

$$[t,G_{B'}]\,\underline{i}^{\#} = \underline{i}\, t^{B''}$$

for every $[t,G_{B'}] \in F(\mathcal{C}(T_2),G_{B'},u)$.

We will prove first that the homomorphism $i^{\#}$ is surjective and second that it is injective, too. By Theorem 2.4.2 then follows that $i^{\#}$ is an isomorphism.

If $b \in B''_s$ and $s \in S(T_1)$ then b itself is a variable of the $S(T_2)$-sorted system of variables u and so

$$[b,G_{B'}]\; i^{\#} = b,$$

i.e. $i^{\#}_s$ is surjective.

If $b \in B''_s \subseteq A''_s$ and $s \in S(T_2)$, $s \notin S(T_1)$ then according to the closure operation (i) there exists a term $t \in T(\Sigma(T_2),w)_s$ with $b = \underline{e}\, t^{A''}$ and $var(t) \subseteq B$ so that

$$b = \underline{e}\, t^{A''} = \underline{i}\, t^{B''} .$$

For the proof of the injectivity let us assume

$$[t_1, G_{B''}]\, i^{\#} = [t_2, G_{B''}]\, i^{\#} .$$

This implies

$$\underline{i}\, t_1^{B''} = \underline{i}\, t_2^{B''} \quad \text{and} \quad \underline{e}\, t_1^{A''} = \underline{e}\, t_2^{A''} .$$

Owing to the closure operation (ii) a finite subset $G_{(w:t_1 \overset{e}{=} t_2)}$ of $G_{A''}$ exists with

$$\mathcal{O}\!\mathcal{l}(T_2) \models (w\colon G_{(w:t_1 \overset{e}{=} t_2)} \longrightarrow t_1 \overset{e}{=} t_2)$$

and $var(G_{(w:t_1 \overset{e}{=} t_2)}) \subseteq B$.

This last relation involves that $G_{(w:t_1 \overset{e}{=} t_2)}$ may be regarded as a set of existence equations on the $S(T_2)$-sorted system of variables u , and so

$$(w{\downarrow}u\colon G_{(w:t_1 \overset{e}{=} t_2)})$$

is a finite subset of $G_{B''}$ and

$$\mathcal{O}\!\mathcal{l}(T_2) \models (w{\downarrow}u\colon G_{(w:t_1 \overset{e}{=} t_2)} \longrightarrow t_1 \overset{e}{=} t_2) .$$

This implies

$$[t_1, G_{B''}] = [t_2, G_{B''}]$$

so that the homomorphism

$$i^{\#} : F(\mathcal{O}\!\mathcal{l}(T_2), G_{B''}, u) \longrightarrow B''$$

is injective.

The cardinality of B_s for every $s \in S(T)$ cannot be greater than the maximum of the cardinality of M_s and $\mathbb{N}$, because $var(t)$ is a finite $S(T)$-sorted set for every term t , $\Omega(T)$ is a finite set of operators, and Δ is a finite set of initial restrictions. ∎

By means of Theorem 4.1.3 we can prove that some classes of algebraic structures cannot be described as a class of models of any canon.

Let us consider for instance the following theory:

SETS <u>is</u> <u>sorts</u> Elements, Sets

<u>oprn</u> $\emptyset \longrightarrow$ Sets

add(Sets,Elements) $\longrightarrow$ Sets

<u>axioms</u> x,y: Elements, s: Sets

add(add(s,x),x) = add(s,x)

add(add(s,x),y) = add(add(s,y),x)

and the class

$$\mathcal{T} = \{ A \in ALG(SETS) \mid A_{Sets} = \text{set of all subsets of } A_{Elements}, \; \emptyset^A = \text{empty subset of } A_{Elements}, \text{ and } add^A(s,x) = s \cup \{x\} \} .$$

Then there is no canon (T,Δ) with $ALG(T,\Delta) = \mathcal{T}$, since for the algebra $A \in \mathcal{T}$, with $A_{Elements} = \mathbb{N}$ and the two-sorted subset M with $M_{Elements} = \mathbb{N}$ and $M_{Sets} = \emptyset$ there is no subalgebra B of A with $B \in \mathcal{T}$ and $M \subseteq B$ so that B_{Sets} is of countable cardinality. $B \in \mathcal{T}$ implies that B_{Sets} is of uncountable cardinality. This example proves that the usual powerset construction cannot be specified by the initial canon specification method.

In the same way one can see that the class of all continuous fields can also not be specified in terms of initial canons. The continuous field of real numbers does not contain a countable continuous subfield. However, the class of all fields can easily be specified by means of a suitable canon.

A further example of a class not specfiable as model class of a canon is the class of all complete lattices. To see this let us consider the complete lattice A of all sets of natural numbers and let us take $M \subseteq A$ as the countable set of all atoms and anti-atoms of A. Once more one can prove that no complete countable sublattice of A exists that includes M.

The following property of model classes of canons is most suitable representable in terms of the theory of categories, see MacLane (1971), namely as closure property with respect to special colimits. Because we do not suppose that readers are familiar with these concepts we will introduce them here .

Let G be any finite directed graph and let (T,Δ) be any canon. A diagram

$$\Phi : G \longrightarrow ALG(T,\Delta)$$

in the class of models of the canon (T,Δ) is given by two functions

$$\Phi_{Nodes} : G_{Nodes} \longrightarrow ALG(T,\Delta)$$

$$\Phi_{Edges} : G_{Edges} \longrightarrow \text{class of all homomorphism between } (T,\Delta)\text{-algebras}$$

so that for every $k \in G_{Edges}$ with $n_1 = k\,begin^G$ and $n_2 = k\,end^G$

$$k\Phi_{Edges} : n_1\Phi_{Nodes} \longrightarrow n_2\Phi_{Nodes}$$

is a homomorphism from the image of n_1 to the image of n_2.

Let $\Phi : G \longrightarrow ALG(T,\Delta)$ be any digram in $ALG(T,\Delta)$. A T-equoid L together with a family

$$(f(n) : n\Phi_{Nodes} \longrightarrow L \mid n \in G_{Nodes})$$

of homomorphisms is called a c o l i m i t of Φ in $ALG(T,\Delta)$ if

(1) $(k\Phi_{Edges})\circ f(n_2) = f(n_1)$ for every edge $k \in G_{Edges}$ directed from the node n_1 to the node n_2 ;

(2) For every family

$$(g(n) : n\Phi_{Nodes} \longrightarrow A \mid n \in G_{Nodes})$$

of homomorphisms with A ALG(T) and

$(k\Phi_{Edges})\circ g(n_2) = g(n_1)$ for every edge $k \in G_{Edges}$ directed from n_1 to n_2 there exists exactly one homomorphism

$$h : L \longrightarrow A$$

with $f(n)\circ h = g(n)$ for every $n \in G_{Nodes}$.

All so-called quotient constructions dealt with in Section 3.3 are constructions of colimits in ALG(T) for special kinds of diagrams. Using the terminology just introduced the results of Section 3.3 can be summarized by the statement that for every finite diagram a colimit exists in ALG(T). It is evident that a colimit is determined uniquely up to isomorphism by the diagram.

We say that a canon model class $ALG(T,\Delta) \subseteq ALG(T)$ is closed with respect to a colimit of a diagram $\Phi : G \longrightarrow ALG(T,\Delta)$ if the existing colimit L of $\Phi : G \longrightarrow ALG(T,\Delta)$ in ALG(T) is even in $ALG(T,\Delta)$.

We say that an initial restriction

$$r = (T_1, \varphi : T_2 \longrightarrow T)$$

in T preserves a colimit $L \in ALG(T,\Delta)$ of $\Phi : G \longrightarrow ALG(T,\Delta)$ if at first $L\varphi$ is a colimit of the diagram

$$\Phi\varphi : G \longrightarrow ALG(T_2)$$

defined by

$$n(\Phi\varphi) = (n\Phi_{Nodes})\varphi \quad \text{for every node } n \in G_{Nodes} ,$$

$$k(\Phi\varphi) = (k\Phi_{Edges})\varphi \quad \text{for every edge } k \in G_{Edges} ,$$

and if secondly

$$(L\varphi){\downarrow}T_1$$

is a colimit in $ALG(T_1)$ of the diagram

$$\Phi\varphi{\downarrow}T_1 : G \longrightarrow ALG(T_1)$$

defined by

$$n(\Phi\varphi{\downarrow}T_1) = ((n\Phi_{Nodes})\varphi){\downarrow}T_1 \quad \text{for every } n \in G_{Nodes} ,$$

$$k(\Phi\varphi{\downarrow}T_1) = ((k\Phi_{Edges})\varphi){\downarrow}T_1 \quad \text{for every } k \in G_{Edges} .$$

Now we are able to state a further closure property of canon model classes.

Theorem 4.1.4. A canon model class $ALG(T,\Delta)$ is closed with respect to those colimits that are preserved by every initial restriction of Δ.

Proof. Let be given a diagram $\Phi : G \longrightarrow ALG(T,\Delta)$ and a colimit

$$(f(n) : n\Phi_{Nodes} \longrightarrow L \mid n \in G_{Nodes})$$

of Φ in $ALG(T)$, and any initial restriction $(T_1, \varphi : T_2 \longrightarrow T)$ of Δ. Then we introduce the following abbreviations:

$$f''(n) = f(n)\varphi$$

$$f'(n) = (f(n))\varphi{\downarrow}T_1$$

for every $n \in G_{Nodes}$.

Now we have to prove that $L\varphi$ is freely generated by $(L\varphi){\downarrow}T_1$, i.e. we have to show that for each $B \in ALG(T_2)$ and every homomorphism

$$h : (L\varphi){\downarrow}T_1 \longrightarrow B{\downarrow}T_1$$

exactly one homomorphism

$$h^{\#} : L\varphi \longrightarrow B$$

with $h = h^{\#}{\downarrow}T_1$ exists.

Because $n\Phi_{Nodes} \in ALG(T,\Delta)$ for every $n \in G_{Nodes}$, the equoid $n(\Phi\varphi) \in ALG(T_2)$ is freely generated by $n(\Phi\varphi{\downarrow}T_1) \in ALG(T_1)$, so that for every $n \in G_{Nodes}$ exactly one homomorphism

$$h'(n) : n(\Phi\varphi) \longrightarrow B$$

exists with $h'(n){\downarrow}T_1 = f''(n)\circ h$.

Since $(T_1, \varphi : T_2 \longrightarrow T)$ preserves the given colimit, there is exactly one homomorphism

$$h^{\#} : L\varphi \longrightarrow B$$

with $f'(n)\circ h^{\#} = h'(n)$ for every $n \in G_{Nodes}$.

It remains to prove that $h^{\#}{\downarrow}T_1 = h$ and that $h^{\#}$ is uniquely determined by this condition.

Using the equalities $f'(n)\circ h^{\#} = h'(n)$, $h'(n){\downarrow}T_1 = f''(n)\circ h$, and $f''(n) = f'(n){\downarrow}T_1$ for every $n \in G_{Nodes}$ we obtain

$$f''(n)\circ(h^{\#}{\downarrow}T_1) = (f'(n){\downarrow}T_1)\circ(h^{\#}{\downarrow}T_1) = (f'(n)\circ h^{\#}){\downarrow}T_1 =$$

$$= h'(n){\downarrow}T_1 = f''(n)\circ h$$

for every $n \in G_{Nodes}$ which implies $h = h^{\#}{\downarrow}T_1$ since

$$(\ f''(n) : n(\Phi\varphi{\downarrow}T_1) \longrightarrow (L\varphi){\downarrow}T_1 \mid n \in G_{Nodes}\ )$$

is a colimit of

$$\Phi\varphi{\downarrow}T_1 : G \longrightarrow ALG(T_1) \ .$$

If $h^{\#\#} : L\varphi \longrightarrow B$ is any homomorphism with $h^{\#\#}{\downarrow}T_1 = h$ we obtain

$$(f'(n)\circ h^{\#\#}){\downarrow}T_1 = f''(n)\circ h$$

for every $n \in G_{Nodes}$ and so

$$h'(n) = f'(n)\circ h^{\#} = f'(n)\circ h^{\#\#}$$

for every $n \in G_{Nodes}$ because $n(\Phi\varphi)$ is freely generated by $n(\Phi\varphi{\downarrow}T_1)$ for every $n \in G_{Nodes}$.

This leads to $h^{\#\#} = h^{\#}$ since

$$(\ f'(n) : n(\Phi\varphi) \longrightarrow L\varphi \mid n \in G_{Nodes}\ )$$

is a colimit of $\Phi\varphi : G \longrightarrow ALG(T_2)$. ∎

If we apply the introduced terminology of colimits to Corollary 3.4.4 we can say that for every theory T and every diagram

$$\Phi : G \longrightarrow ALG(T)$$

where G is a not necessarily finite chain, the $S(T)$-carrier of a colimit of Φ in $ALG(T)$ is a colimit of the diagram

$$\Phi{\downarrow}T_o : G \longrightarrow ALG(T_o)$$

where T_o denotes the subtheory of T with $S(T_o) = S(T)$, $\Omega(T_o) = \emptyset$ and $\mathcal{Q}(T_o) = \emptyset$, i.e. $ALG(T_o)$ is the class of all

S(T)-sets.
An immediate consequence of this fact is the following preservation property of theory-morphisms.

Corollary 4.1.5. Let $\varphi : T_1 \longrightarrow T_2$ be any theory-morphism, $\Phi : G \longrightarrow ALG(T_2)$ may be a chain of T_2-algebras, and

$$(f(n) : n\Phi \longrightarrow L \mid n \in G_{Nodes})$$

may be a colimit of the diagram Φ in $ALG(T_2)$. Then

$$(f(n) : n(\Phi\varphi) \longrightarrow L\varphi \mid n \in G_{Nodes})$$

is a colimit of the diagram

$$\Phi\varphi : G \longrightarrow ALG(T_1)$$

in $ALG(T_1)$. ∎

Without any additional problems Corollary 4.1.5 can be proved for so-called directed diagrams $\Phi : G \longrightarrow ALG(T_2)$, where a diagram $\Phi : G \longrightarrow ALG(T_2)$ is called directed if

(1) $x\,begin^G = y\,begin^G$ and $x\,end^G = y\,end^G$ implies $x = y$ for all $x, y \in G_{Edges}$;

(2) If $x\,begin^G = y\,begin^G$ for $x, y \in G_{Edges}$ then there are $x^\#, y^\# \in G_{Edges}$ with $x\,end^G = x^\#\,begin^G$, $y\,end^G = y^\#\,begin^G$ and $x^\#\,end^G = y^\#\,end^G$;

(3) If $x\,end^G = y\,begin^G$ for $x, y \in G_{Edges}$ then there is an edge $z \in G_{Edges}$ with $x\,begin^G = z\,begin^G$ and $y\,end^G = z\,end^G$.

A combination of Theorem 4.1.4 with this extended version of Corollary 4.1.5 yields the following closure property.

Theorem 4.1.6. Any canon model class $ALG(T,\Delta) \subseteq ALG(T)$ is closed in $ALG(T)$ with respect to colimits of directed diagrams in $ALG(T,\Delta)$.

4.2 Computability and Definability on parameterized abstract data types

In Chapter 1 the usefulness of functional enrichments of specifications is demonstrated by several examples. In this section we investigate the question if a new data type or a new operation can be reduced effectively to the given data types and given operations.

For the sake of conceptual simplicity a specification should distinguish explicitly a minimal set of fundamental operations and should characterize the other ones as computable derived from the fundamental operations. This requires a mathematically precise notion of a computable extension of a specification. As announced earlier we prove in this section that the concept of structural induction on partial algebras, or in other words the concept of initial restrictions in algebraic theories with partial operations, is sufficient for the definition of computable extensions of specifications. This purely algebraic version of parameterized algorithms was suggested by H.Kaphengst in Kaphengst, Reichel (1971) and improved in Kaphengst (1981). The concept of a definable extension of a specification was introduced by K.Benecke in his thesis in which he also proved some necessary conditions for computable and uniquely specifiable extensions, Theorem 4.2.4 and Theorem 4.2.8 below.

Roughly speaking, a parameterized algorithm $\mathcal{A}$ associates uniformly with every algebra $A \in ALG(T,\Delta)$ of a canon model class a function $A\mathcal{A}$ in such a way that $A\mathcal{A}$ is a partial recursive function provided A is a computable algebra, i.e. the carrier sets A_s, $s \in S(T)$, are recursively enumerable sets and the fundamental operations σ^A, $\sigma \in \Omega(T)$, are partial recursive functions. Analogously, a parameterized computable set construction $\mathcal{N}$ associates uniformly with every algebra $A \in ALG(T,\Delta)$ a recursively enumerable set $A\mathcal{N}$ provided A is a computable algebra.

Since the parameterization is always represented by a class of models of a canon (T,Δ) we use the notions (T,Δ)-algorithm and (T,Δ)-reset (recursively enumerable set).

Definition 4.2.1. Let (T,Δ) be any canon. A correspondence $\mathcal{N}$ that associates with every algebra $A \in ALG(T,\Delta)$ a set $A\mathcal{N}$ is called a (T,Δ) - r e s e t , if there is a finite algebraic theory $T^{\#}$ including the algebraic theory T, a finite S(T)-sorted system of variables $w: X \longrightarrow S(T)$ and a finite set $(w:G)$ of existence equations on $w: X \longrightarrow S(T)$ with operators of $\Omega(T^{\#})$ such that

(1) for every $A \in ALG(T,\Delta)$ there exists a $T^{\#}$-algebra $A^{\#}$ with $A^{\#}{\downarrow}T = A$ which is freely generated by A ;

(2) $A\mathcal{N} = A^{\#}_{(w:G)}$ for every $A \in ALG(T,\Delta)$ where $A^{\#}$ is used in the sense of (1).

$(T^{\#},G,w)$ is called a r e p r e s e n t a t i o n of the (T,Δ)-reset . ∎

Definition 4.2.2. Let (T,Δ) be any canon. A correspondence $\mathcal{A}$ that associates with every algebra $A \in ALG(T,\Delta)$ a function $A\mathcal{A}$ is said to be a (T,Δ) - *a l g o r i t h m* , if there is a finite algebraic theory $T^{\#}$ including the algebraic theory T and an operator $\sigma \in \Omega(T^{\#})$ such that

(1) $\sigma : w \longrightarrow s$ in $\Omega(T^{\#})$ with $w : [n] \longrightarrow S(T)$, $s \in S(T)$;

(2) every $A \in ALG(T,\Delta)$ generates freely a $T^{\#}$-algebra $A^{\#}$ with $A^{\#}\downarrow T = A$;

(3) $A\mathcal{A} = \sigma^{A^{\#}}$ for every $A \in ALG(T,\Delta)$ where $A^{\#}$ is used in the sense of (2).

$(T^{\#},\sigma)$ is called a *r e p r e s e n t a t i o n* of the (T,Δ)-algorithm $\mathcal{A}$. ∎

It is evident that for every (T,Δ)-algorithm $\mathcal{A}$ represented by $(T^{\#},\sigma)$ with $\sigma : w \longrightarrow s$ the triple $(T^{\#}, \mathrm{def}\,\sigma, w)$ is a representation of a (T,Δ)-reset.

Examples of the notions introduced in the preceding definitions can easily be taken from the specifications of Chapter 1 .

However, we suppose that these examples do not justify sufficiently the preceding definitions. Therefore we will give a further justification by proving that the introduced concepts coincide with the classical ones in the case $(T,\Delta) = \mathbb{C}_{NAT_0}$ with

NAT_0 __is__ __definition__
  __sorts__ N
  __oprn__ $0 \longrightarrow N$
    $s(N) \longrightarrow N$
  __axioms__ none
__end__ NAT_0

Theorem 4.2.3. The NAT_0-algorithms are precisely the partial recursive functions on natural numbers, and the NAT_0-resets are precisely the recursively enumerable sets of n-tuples of natural numbers.

__Proof__. Without loss of generality we may assume that

$$\mathbb{N} = T(\Sigma(NAT_0), \lambda)_N .$$

At first we prove that every NAT_0-reset is recursively enumerable. Therefore let $(T^{\#}, G, w)$ with $w : [n] \longrightarrow N$ be any representation of a NAT_0-reset.

If $A^{\#} \in ALG(T^{\#})$ is freely generated by $\mathbb{N}$ and $A^{\#} \downarrow T(NAT_0) = \mathbb{N}$, then $A^{\#}$ is isomorphic to $F(\mathcal{Q}(T^{\#}), \emptyset, \lambda)$ according to the proof of Theorem 3.5.6 and Corollary 3.5.7.
According to Theorem 3.2.3

$$\underline{t} \in \mathbb{N}_w = A^{\#}_w \subseteq T(\Sigma(T^{\#}), \lambda)_w$$

is a solution of $(w:G)$ in $A^{\#}$ if and only if

$$[\underline{t}, \emptyset] \in F(\mathcal{Q}(T^{\#}), \emptyset, \lambda)_{(w:G)},$$

i.e. iff

$$\mathcal{Q}(T^{\#}) \models (\lambda : \emptyset \longrightarrow G\tilde{\underline{t}}) .$$

The results of Section 3.1 imply that the set of all finite sets of existence equations $(\lambda:H)$, with operators taken from $\Omega(T^{\#})$, satisfying

$$\mathcal{Q}(T^{\#}) \models (\lambda : \emptyset \longrightarrow H)$$

is recursively enumerable.
The function that associates with every term $\underline{t} \in \mathbb{N}_w$ the finite set of existence equations $(\lambda : G\tilde{\underline{t}})$ is evidently a computable function, and so $A^{\#}_{(w:G)}$ is representable as inverse image of a recursively enumerable set with respect to the computable function just described. This proves the recursive enumerability of $A^{\#}_{(w:G)}$.

As next we prove that every NAT_0-algorithm is a computable function. Let $(T^{\#}, \sigma)$ be any representation of a NAT_0-algorithm. As above we use the $T^{\#}$-algebra $F(\mathcal{Q}(T^{\#}), \emptyset, \lambda)$ and apply the results of Section 3.1. As a consequence we can state that the set

$$R = \{ (t_1, t_2) \in T(\Sigma(T^{\#}), \lambda) \times T(\Sigma(NAT_0), \lambda) \mid \mathcal{Q}(T^{\#}) \models (\lambda : \emptyset \longrightarrow t_1 \stackrel{e}{=} t_2) \}$$

is recursively enumerable.
Because of

$$F(\mathcal{Q}(T^{\#}), \emptyset, \lambda) \downarrow T(NAT_0) \xrightarrow{\sim} T(\Sigma(NAT_0), \lambda) = \mathbb{N}$$

this set R is the graph of a partial recursive function f_R.
If

$$\underline{t} \in \operatorname{dom} \sigma^{A^{\#}}$$

then we obtain

$$\mathcal{Q}(T^{\#}) \models (\lambda : \emptyset \longrightarrow \underline{t}\,\sigma^{T(\Sigma(T^{\#}), \lambda)} \stackrel{e}{=} \underline{t}\,\sigma^{T(\Sigma(T^{\#}), \lambda)})$$

and

$$\underline{t}\,\sigma^{A^{\#}} = (\underline{t}\,\sigma^{T(\Sigma(T^{\#}), \lambda)})\, f_R \in T(\Sigma(NAT_0), \lambda).$$

According to the first part of this proof $\sigma^{A^{\#}}$ is the restriction of the composition

$$\sigma^{T(\Sigma(T^{\#}),\lambda)} \circ f_R$$

of two computable functions to the recursively enumerable set

$$\mathrm{dom}\,\sigma^{A^{\#}}$$

and so $\sigma^{A^{\#}}$ itself is partial recursive.

Up to now we proved the compatibility of the introduced concepts with the classical ones. It remains the proof that they are strong enough. We complete the proof by showing that the recursive functionals of superposition, primitive recursion and minimization, which generate the set of partial recursive functions starting with the constant zero, the successor function and the projections, can be expressed by the introduced concepts.

By Definition 4.2.2 it is evident that the constant function zero, the successor function and the projections are NAT_0-algorithms, and that the superposition of NAT_0-algorithms is also a NAT_0-algorithm.

Due to the Theorem of Kleene, see Rogers Jr. (1967), it suffices to show that for all total NAT_0-algorithms

$$g: \mathbb{N}^k \longrightarrow \mathbb{N}, \quad h: \mathbb{N}^{k+2} \longrightarrow \mathbb{N}$$

the function

$$r(g,h): \mathbb{N}^{k+1} \longrightarrow \mathbb{N}$$

defined by the functional of primitive recursion is a NAT_0-algorithm, and that for every total NAT_0-algorithm

$$g: \mathbb{N}^{k+1} \longrightarrow \mathbb{N}$$

the partial function

$$\min_y((x_0,\dots,x_{k-1},y)\, g = 0)$$

defined by minimization, or by the μ-operator, is also a NAT_0-algorithm.

Let $(T_1^{\#},\sigma_1)$, $(T_2^{\#},\sigma_2)$ be representations of $g: \mathbb{N}^k \longrightarrow \mathbb{N}$ and $k: \mathbb{N}^{k+2} \longrightarrow \mathbb{N}$, respectively. Then there exists a finite extension $T^{\#}$ of $T(NAT_0)$ containing σ_1 and σ_2 in such a way that $(T^{\#},\sigma_1)$, $(T^{\#},\sigma_2)$ is a representation of the function g, h, respectively.
Now we set

$T^{\#\#}$ __is__ $T^{\#}$ __with__

  __oprn__ $r(N^{k+1}) \longrightarrow N$

  __axioms__ $x_o, x_1, \ldots, x_k : N$

  $r(x_o, x_1, \ldots, x_{k-1}, 0) = \mathfrak{G}_1(x_o, \ldots, x_{k-1})$

  $r(x_o, x_1, \ldots, x_{k-1}, s(x_k)) = \mathfrak{G}_2(x_o, \ldots, x_{k-1}, r(x_o, \ldots, x_k), x_k)$

__end__ $T^{\#\#}$

and we obtain a representation $(T^{\#\#}, r)$ of $r(g,h)$.

Now let $(T^{\#}, \varrho)$ be a representation of a NAT_o-algorithm $g: \mathbb{N}^{k+1} \longrightarrow \mathbb{N}$. For the specification of the μ-operator we need some auxiliary NAT_o-algorithms. For that reason we begin with

$T_1^{\#}$ __is__ $T^{\#}$ __with__

  __oprn__ $l(N,N) \longrightarrow N$

  $eq(N,N) \longrightarrow N$

  __axioms__ $x, x_1, x_2 : N$

  $l(0,x) = 0$

  __if__ $l(x_1,x_2) = 0$ __then__ $l(s(x_1), s(x_2)) = 0$

  $l(s(x),0) = s(o)$

  __if__ $l(x_1,x_2) = s(0)$ __then__ $l(s(x_1), s(x_2)) = s(0)$

  __if__ $l(x_1,x_2) = 0$, $l(x_2,x_1) = 0$ __then__ $eq(x_1,x_2) = 0$

  __if__ $l(x_1,x_2) = s(0)$ __then__ $eq(x_1,x_2) = s(0)$

  __if__ $eq(x_1,x_2) = s(0)$ then $eq(x_2,x_1) = s(0)$

__end__ $T_1^{\#}$.

$(T_1^{\#}, l)$, $(T_1^{\#}, eq)$ are representations of the NAT_o-algorithms $l' : \mathbb{N} \times \mathbb{N} \longrightarrow \mathbb{N}$, $eq' : \mathbb{N} \times \mathbb{N} \longrightarrow \mathbb{N}$, respectively, with

$l'(x,y) = 0$ iff $x \leq y$ for all $x,y \in \mathbb{N}$,

$l'(x,y) = 1$ iff $x \nleq y$ for all $x,y \in \mathbb{N}$,

$eq'(x,y) = 0$ iff $x = y$ for all $x,y \in \mathbb{N}$,

$eq'(x,y) = 1$ iff $x \neq y$ for all $x,y \in \mathbb{N}$.

In the next step we set

$T_2^{\#}$ __is__ $T_1^{\#}$ __with__

  __sorts__ P

  __oprn__ $e \longrightarrow P$

  $v(N,P) \longrightarrow P$

  $min(P) \longrightarrow N$

  __axioms__ $x, x_1, x_2 : N$, $p : P$

$v(x_1, v(x_2, p)) = v(x_2, v(x_1, p))$
$v(x, v(x, p)) = v(x, p)$
$min(e) = 0$
<u>if</u> $l(x, min(p)) = 0$ <u>then</u> $min(v(x, p)) = x$
<u>if</u> $l(x, min(p)) = s(0)$ <u>then</u> $min(v(x, p)) = min(p)$
<u>end</u> $T_2^\#$.

If $A^\# \in ALG(T_2^\#)$ is freely generated by $\mathbb{N}$, then $A^\# \downarrow T(NAT_o) = \mathbb{N}$ and $A_P^\#$ is up to isomorphism the set of all finite subsets of $\mathbb{N}$ with $e^{A^\#} = \emptyset$ and

$$v^{A^\#}(x, p) = p \cup \{x\}$$

for every $p \in A_P^\#$, $x \in \mathbb{N}$, and

$$min^{A^\#} : A_P^\# \longrightarrow \mathbb{N}$$

associates with every finite set of natural numbers its minimal element with respect to the natural ordering.

The specification of the μ-operator is finally given by

$T_3^\#$ <u>is</u> $T_2^\#$ <u>with</u>
<u>sorts</u> Q
<u>oprn</u> $q_o \longrightarrow Q$
$q(N^{k+1}) \longrightarrow Q$
$d(N^{k+1}) \longrightarrow P$
$m(x_o{:}N, \ldots, x_{k-1}{:}N$ <u>iff</u> $q(x_o, \ldots, x_{k-1}, 0) = q_o) \longrightarrow N$
<u>axioms</u> $x_o, x_1, \ldots, x_k, x_{k+1} : N$
<u>if</u> $\varrho(x_o, \ldots, x_{k-1}, x_k) = 0,\ l(x_{k+1}, x_k) = 0$
<u>then</u> $q(x_o, \ldots, x_{k-1}, x_{k+1}) = q_o$ <u>fi</u>
<u>if</u> $\varrho(x_o, \ldots, x_{k-1}, 0) = 0$ <u>then</u> $d(x_o, \ldots, x_{k-1}, 0) = v(0, e)$
<u>if</u> $eq(o(x_o, \ldots, x_{k-1}, 0), 0) = s(0)$ <u>then</u> $d(x_o, \ldots, x_{k-1}, 0) = e$
<u>if</u> $\varrho(x_o, \ldots, x_{k-1}, s(x_k)) = s(0)$ <u>then</u>
$d(x_o, \ldots, x_{k-1}, s(x_k)) = v(s(x_k), d(x_o, \ldots, x_k))$ <u>fi</u>
<u>if</u> $eq(o(x_o, \ldots, x_{k-1}, s(x_k)), 0) = s(0)$ <u>then</u>
$d(x_o, \ldots, x_{k-1}, s(x_k)) = d(x_o, \ldots, x_k)$ <u>fi</u>
<u>if</u> $\varrho(x_o, \ldots, x_{k-1}, x_k) = 0$ <u>then</u>
$m(x_o, \ldots, x_{k-1}) = min(d(x_o, \ldots, x_k))$ <u>fi</u>
<u>end</u> $T_3^\#$.

Up to isomorphism an algebra $A^{\#} \in ALG(T_3^{\#})$ freely generated by $\mathbb{N}$ is characterized by

$A^{\#} \downarrow T(NAT_o) = \mathbb{N}$,

$A_P^{\#}$ = set of all finite subsets of $\mathbb{N}$,

$A_Q^{\#} = \{q_o\} \cup \{(x_o,\ldots,x_k) \in \mathbb{N}^{k+1} \mid (x_o,\ldots,x_{k-1},y)\, g = 0$ for all $y \le x_k\}$,

$e^{A^{\#}} = \emptyset$,

$(x,y)\, v^{A^{\#}} = \{x\} \cup y$ for $x \in \mathbb{N}$, $y \in A_P^{\#}$,

$y\, min^{A^{\#}} = min\{x \in \mathbb{N} \mid x \in y\}$ for all $y \in A_P^{\#}$ with $y \neq \emptyset$,

$\emptyset\, min^{A^{\#}} = 0$,

$q_o^{A^{\#}} = q_o$,

$$(x_o,\ldots,x_k)\, q^{A^{\#}} = \begin{cases} q_o & \text{if there is at least one } y \in \mathbb{N} \text{ with } x_k \le y \text{ and } (x_o,\ldots,x_{k-1},y)\, g = 0 \\ (x_o,\ldots,x_k) & \text{else} \end{cases}$$

$(x_o,\ldots,x_k)\, d^{A^{\#}} = \{x \in \mathbb{N} \mid x \le x_k ,\ (x_o,\ldots,x_{k-1},x)\, g = 0\}$,

$(x_o,\ldots,x_{k-1})\, m^{A^{\#}}$ is defined if and only if there is a $y \in \mathbb{N}$ with $(x_o,\ldots,x_{k-1},y)\, g = 0$,

i.e. iff $(x_o,\ldots,x_{k-1},0)\, q^{A^{\#}} = q_o$, if

$(x_o,\ldots,x_{k-1})\, m^{A^{\#}}$ is defined, then

$(x_o,\ldots,x_{k-1})\, m^{A^{\#}}$ = smallest natural number y with $(x_o,\ldots,x_{k-1},y)\, g = 0$.

This proves that $(T_3^{\#},m)$ is a representation of the NAT_o-algorithm defined by the μ-operator

$min_y((x_o,\ldots,x_{k-1},y)\, g = 0)$. ∎

It was not our aim to give the shortest proof of the justification theorem. A side effect of this proof should be a further illustration of the developed approach. Therefore we split up the specification of the μ-operator into isolated modules.

We will not discuss the justification of the introduced concepts in all details. A more profound result in this direction can be found in Kaphengst (1981).

We resume the problem of deciding if an intended functional enrichment of a specification can be specified by an enrichment by definition or not. A first result in this direction is the following property of (T,Δ)-algorithms and (T,Δ)-resets.

Theorem 4.2.4. (T,Δ)-algorithms and (T,Δ)-resets are compatible with homomorphisms between (T,Δ)-algebras, i.e. if $\mathcal{A}$ is any (T,Δ)-algorithm with

$$A\mathcal{A}: A_w \longrightarrow A_s \ , \ w: X \longrightarrow S(T), \ s \in S(T)$$

for every $A \in ALG(T,\Delta)$, and if $\mathcal{N}$ is any (T,Δ)-reset with

$$A\mathcal{N} \subseteq A_v \ , \ v: Y \longrightarrow S(T)$$

for every $A \in ALG(T,\Delta)$, then the following conditions are satisfied for every homomorphism $f: A \longrightarrow B$ with $A, B \in ALG(T,\Delta)$.

(1) $\underline{a} \in dom(A\mathcal{A}) \subseteq A_w$ implies $\underline{a} f_w \in dom(B\mathcal{A})$ and
$(\underline{a}(A\mathcal{A}))f_s = (\underline{a} f_w)(B\mathcal{A})$.

(2) $\underline{a} \in A\mathcal{N} \subseteq A_v$ implies $\underline{a} f_v \in B\mathcal{N}$.

Proof. The compatibility with homomorphisms between (T,Δ)-algebras is an immediate consequence of the definition.
Let $(T^{\#}, \sigma)$ be a representation of $\mathcal{A}$, then

$$A\mathcal{A} = \sigma^F, \ B\mathcal{A} = \sigma^{F'}, \ A = F{\downarrow}T, \ B = F'{\downarrow}T$$

if $F, F' \in ALG(T^{\#})$ are freely generated by A, B, respectively. For the homomorphism $f: A \longrightarrow B$ exists exactly one homomorphism $f^{\#}: F \longrightarrow F'$ with $f^{\#}{\downarrow}T = f$ and so $f_w = f^{\#}_w$ for $w: X \longrightarrow S(T)$.
Since $f^{\#}$ is a homomorphism we obtain condition (1) above.
Since homomorphisms preserve solutions of sets of existence equations condition (2) can be proved in the same way. ∎

Even if Theorem 4.2.4 can be proved easily it seems to be of some practical value as it was demonstrated by its application in Chapter 1.

The concept of canons makes available not only a formalization of computable extensions of specifications. In the same way as first order theories can be extended by definition canons can be extended.

Definition 4.2.5. Let (T,Δ) be any canon. A correspondence $\mathcal{A}$ that associates with every $A \in ALG(T,\Delta)$ a mapping $A\mathcal{A}$ is called a d e f i n a b l e (T,Δ) - m a p p i n g if there is a finite extension $T^{\#}$ of the algebraic theory T with $\sigma \in \Omega(T^{\#})$ such that

(1) $\sigma: w \longrightarrow s$, $w: [n] \longrightarrow S(T)$, $s \in S(T)$;

(2) For every $A \in ALG(T,\Delta)$ there exists at least one $A^{\#} \in ALG(T^{\#})$ with $A^{\#}{\downarrow}T = A$;

(3) For all $A^{\#}, B^{\#} \in ALG(T^{\#})$ with $A^{\#}\downarrow T = B^{\#}\downarrow T = A \in ALG(T,\Delta)$ follows $\delta^{A^{\#}} = \delta^{B^{\#}} = A\alpha$.

$(T^{\#}, \delta)$ is called a r e p r e s e n t a t i o n of the definable (T,Δ)-mapping α.

A correspondence $\mathcal{N}$ that associates with every $A \in ALG(T,\Delta)$ a set $A\mathcal{N}$ is said to be a d e f i n a b l e (T,Δ)-s e t if there is a finite extension $T^{\#}$ of the algebraic theory T, a finite $S(T)$-sorted system of variables $w: X \longrightarrow S(T)$ and a finite set of existence equations $(w:G)$ using operators of $\Omega(T^{\#})$ such that

(1) For every $A \in ALG(T,\Delta)$ there is at least one $A^{\#} \in ALG(T,\Delta)$ with $A^{\#}\downarrow T = A$;

(2) For all $A^{\#}, B^{\#} \in ALG(T,\Delta)$ with $A^{\#}\downarrow T = B^{\#}\downarrow T = A \in ALG(T,\Delta)$ follows $A^{\#}_{(w:G)} = B^{\#}_{(w:G)} = A\mathcal{N}$.

$(T^{\#}, G, w)$ is called a r e p r e s e n t a t i o n of the definable (T,Δ)-set . ■

These two different kinds of extensions give rise to the study of the relations between the concepts of computability and definability over canons.

By similar ideas as used in the proof of Theorem 4.2.3 one can prove that over the totally restricted canon NAT_o every NAT_o-algorithm and every NAT_o-reset is a definable NAT_o-mapping and a definable NAT_o-set, respectively. However, in general both concepts are independent.

Corollary 4.2.6. Computability and definability over general canons are incomparable concepts.

Proof. We prove this corollary by giving a (T_1, Δ_1)-reset not definable and by a definable (T_2, Δ_2)-mapping that is not a (T_2, Δ_2)-algorithm.

Let the canon (T_1, Δ_1) be given by $T_1 = T(NAT_o)$ and $\Delta_1 = \emptyset$, so that $ALG(T_1, \Delta_1) = ALG(T(NAT_o))$. For this very simple algebraic theory we consider the (T_1, Δ_1)-reset $\mathcal{N}$ that associates with every $A \in ALG(T(NAT_o))$ the set

$$A\mathcal{N} = \{ x \in A_N \mid x = 0^A \underbrace{s^A \ldots s^A}_{n\text{-times}} ,\ n \in \mathbb{N} \}$$

of all those elements $x \in A_N$ reachable from 0^A by finitely many

applications of $s^A : A_N \longrightarrow A_N$.
A representation of this (T_1, Δ_1)-reset is given by

$$(T_1^{\#},\ q(x)=q_o\ ,\ N\ )$$

with

$T_1^{\#}$ **is** $T(NAT_o)$ **with**
sorts Q
oprn $q_o \longrightarrow Q$
$q(N) \longrightarrow Q$
axioms $x: N$
$q(0) = q_o$
if $q(x) = q_o$ **then** $q(s(x)) = q_o$
end $T_1^{\#}$.

If we assume that this $(T(NAT_o),\emptyset)$-reset is definable and that (T_3,G,N) is a representation of it, then for every $A \in ALG(T_3)$

$$A_{(N:G)} = (A{\downarrow}T(NAT_o))\mathscr{N}$$

where $\mathscr{N}$ denotes the $(T(NAT_o)\emptyset)$-reset represented by $(T_1^{\#},\ q(x)=q_o\ ,\ N\ )$. By adding each existence equation of $(N:G)$ as an axiom to the set $\mathcal{A}(T_3)$ of axioms of T_3 we obtain the theory $T_3^{\#}$. Now for every $B \in ALG(T_3^{\#})$ follows

$$B_N = B_{(N:G)} = (B{\downarrow}T(NAT_o))\mathscr{N}$$

and so the set B_N is at most countable for every $B \in ALG(T_3^{\#})$. However, this property contradicts the following property of every class of models of a theory T : If there is an algebra $A \in ALG(T)$ where A_s consists of at least two elements for some $s \in S(T)$ then for every cardinality α there exists an algebra $B \in ALG(T)$ with $\alpha \leq card\, B_s$. This property can be proved as follows:
Let M be a set with $card\, M = \alpha$ and $w: M \longrightarrow S(T)$ an $S(T)$-sorted system of variables with $xw = s$ for every $x \in M$. Now we consider the T-algebra $F(\mathcal{A}(T),\emptyset,w)$ and the universal solution

$$\underline{e} \in F(\mathcal{A}(T),\emptyset,w)_{(w:\emptyset)} = F(\mathcal{A}(T),\emptyset,w)_w\ .$$

The property is proved if we can show that $\underline{e}$ is an injective assignment. Let $x_1,x_2 \in M$ be any two variables. Then there is an assignment $\underline{f} \in A_w$ with $x_1\underline{f} = x_2\underline{f}$ because $card\, A_s \geq 2$. This implies the existence of a homomorphism

$$f^{\#}: F(\mathcal{A}(T),\emptyset,w) \longrightarrow A$$

with $x_1\underline{e}\,f_s^{\#} = x_1\underline{f}$, $x_2\underline{e}\,f_s^{\#} = x_2\underline{f}$ and so $x_1\underline{e} \neq x_2\underline{e}$ because of $x_1\underline{f} \neq x_2\underline{f}$.

For the second part of the proof let the canon (T_2, Δ_2) be given by

```
(T₂,Δ₂) is definition
        sorts Bool
        oprn  true  ⟶ Bool
              false ⟶ Bool
        axioms  none
        with requirement
        sorts   M
        oprn   none
        axioms  none
    end (T₂, Δ₂).
```

The identity of elements of sort M is a definable (T_2, Δ_2)-mapping represented by $(T_2^{\#}, id)$ with

```
T₂# is T₂ with
        oprn  id(M,M) ⟶ Bool
        axioms  x,y: M
        id(x,x) = true
        if  id(x,y) = true  then  x = y
    end T₂# .
```

By means of Theorem 4.2.4 one can easily prove that this definable (T_2, Δ_2)-mapping is not a (T_2, Δ_2)-algorithm because it is evidently not compatible with every homomorphism between (T_2, Δ_2)-algebras. ∎

One should take note of the fact that definability over canons is not explicitly indicated within specifications. The keyword 'definition' does not indicate the specification of a definable (T, Δ)-set or of a definable (T, Δ)-mapping, on the contrary it indicates usually the specification of a (T, Δ)-algorithm or of a (T, Δ)-reset. Definability has to be expressed by enrichments by requirements.

In practical applications an intended (T, Δ)-mapping is very often specified by alternating enrichments by definition and requirement. A typical example for that case is the specification of the predicate on directed graphs indicating the existence of a finite directed path from one node to another one as specified in C-D-GRAPH, see page 40 and 41. This gives rise to the following definition.

Definition 4.2.7. Let (T, Δ) be any canon. A correspondence $\mathfrak{A}$ that associates with every $A \in ALG(T, \Delta)$ a mapping

$$A\wp : dom(A\wp) \longrightarrow A_s$$

with $dom(A\wp) \subseteq A_w$, $w:[n] \longrightarrow S(T)$, $s \in S(T)$, is called a u n i q u e l y s p e c i f i a b l e (T,Δ) - m a p p i n g if there is a canon $(T^{\#}, \Delta^{\#})$ with $T \subseteq T^{\#}$, $\Delta \subseteq \Delta^{\#}$ and an operator $\sigma \in \Omega(T^{\#})$ such that

(1) For every $A \in ALG(T,\Delta)$ there exists at least one algebra $A^{\#} \in ALG(T^{\#}, \Delta^{\#})$ with $A = A^{\#}{\downarrow}T$;
(2) For all $A^{\#}, B^{\#} \in ALG(T^{\#}, \Delta^{\#})$ with $A^{\#}{\downarrow}T = B^{\#}{\downarrow}T = A \in ALG(T,\Delta)$ it follows $\sigma^{A^{\#}} = \sigma^{B^{\#}} = A\wp$.

$(T^{\#}, \Delta^{\#}, \sigma)$ is called a representation of the uniquely specifiable (T,Δ)-mapping $\wp$.

A correspondence $\mathcal{M}$ that associates with every $A \in ALG(T,\Delta)$ a set $A\mathcal{M} \subseteq A_w$ with $w:[n] \longrightarrow S(T)$, $s \in S(T)$ is called a u n i q u e l y s p e c i f i a b l e (T,Δ) - s e t if there is a canon $(T^{\#}, \Delta^{\#})$ with $T \subseteq T^{\#}$, $\Delta \subseteq \Delta^{\#}$ and a finite set of existence equations $(w:G)$ using operators of $\Omega(T^{\#})$ such that

(1) For every $A \in ALG(T,\Delta)$ there exists at least one algebra $A^{\#} \in ALG(T^{\#}, \Delta^{\#})$ with $A = A^{\#}{\downarrow}T$;
(2) For all $A^{\#}, B^{\#} \in ALG(T^{\#}, \Delta^{\#})$ with $A^{\#}{\downarrow}T = B^{\#}{\downarrow}T = A \in ALG(T,\Delta)$ it follows $A\mathcal{M} = A^{\#}_{(w:G)} = B^{\#}_{(w:G)}$.

$(T^{\#}, \Delta^{\#}, G, w)$ is called a representation of the uniquely specifiable (T,Δ)-set $\mathcal{M}$. ∎

By a slight modification of the preceding example of a (T_1, Δ_1)-reset that is not definable, see proof of Corollary 4.2.6 , we can give in the following an example of a uniquely specifiable (T,Δ)-mapping neither definable nor computable. This example proves that Definition 4.2.7 represents a proper generalization.

The correspondence $\wp$ may associate with every algebra $A \in ALG(T(NAT_o),\emptyset)$ the predicate

$$A\wp : A_N \longrightarrow A_{Bool}$$

with $x(A\wp) = true$ if x is derivable from 0^A by a finite number of applications of $\sigma^A : A_N \longrightarrow A_N$ and $x(A\wp) = false$ otherwise. The specification of this predicate is given by the following canon DERIVABLE:

DERIVABLE <u>is</u> NAT_o <u>with definition</u>
<u>sorts</u> Q

oprn $q_0 \longrightarrow Q$

$q(N) \longrightarrow Q$

axioms $x: N$

$q(0) = q_0$

if $q(x) = q_0$ **then** $q(s(x)) = q_0$

with requirement

oprn $p(N) \longrightarrow Bool$

axioms $x: N$

if $q(x) = q_0$ **then** $p(x) = true$

if $p(x) = true$ **then** $q(x) = q_0$

end DERIVABLE

With this specification (DERIVABLE,p) is a representation of the described $(T(NAT_0),\emptyset)$-predicate.
Since $ALG(T(NAT_0),\emptyset)$ is a model class of a first order theory similar considerations as above enable us to see that this $(T(NAT_0),\emptyset)$-predicate is not a definable $(T(NAT_0),\emptyset)$-mapping.
Theorem 4.2.4 implies that $\mathscr{A}$ is not a $(T(NAT\ ),\emptyset)$-algorithm provided one can define a homomorphism $f: A \longrightarrow B$ with A and B in $ALG(T(NAT_0),\emptyset)$ which is not compatible with $\mathscr{A}$. Such a homomorphism is given by

$A_N = \mathbb{N}$, $0^A = 0$, $x\sigma^A = x+2$ for every $x \in \mathbb{N}$,

$B_N = \mathbb{N}$, $0^B = 0$, $x\sigma^B = x+1$ for every $x \in \mathbb{N}$,

$(2x) f_N = x$, $(2x+1) f_N = x$ for every $x \in \mathbb{N}$, and

$x f_{Bool} = x$ for $x \in \{true, false\}$

because

$3(A\mathscr{A}) f_{Bool} = false$, $3 f_N(B\mathscr{A}) = 1(B\mathscr{A}) = true$, i.e.

$3(A\mathscr{A}) f_{Bool} \neq 3 f_N(B\mathscr{A})$.

We finish this section with a property of uniquely specifiable (T,Δ)-mappings and uniquely specifiable (T,Δ)-sets.

<u>Theorem 4.2.8</u>. Uniquely specifiable (T,Δ)-mappings and uniquely specifiable (T,Δ)-sets are compatible with isomorphisms between (T,Δ)-algebras.

<u>Proof</u>. Let $\mathscr{A}$ be any uniquely specifiable (T,Δ)-mapping with

$A\mathscr{A} : A_w \longrightarrow A_s$, $w: [n] \longrightarrow S(T)$, $s \in S(T)$

and let $\mathscr{S}$ be any uniquely specifiable (T,Δ)-set with

$A\mathscr{S} \subseteq A_v$, $v: [m] \longrightarrow S(T)$

for every $A \in ALG(T,\Delta)$. Additionally let $f: A \longrightarrow B$ be an isomorphism with $A, B \in ALG(T,\Delta)$. According to Definition 4.2.7 there are representations $(T^{\#}, \Delta^{\#}, \delta)$ and $(T^{\#}, \Delta^{\#}, G, w)$ of α and ν, respectively, so that $A^{\#}, B^{\#} \in ALG(T^{\#}, \Delta^{\#})$ exist with $A = A^{\#}{\downarrow}T$, $B = B^{\#}{\downarrow}T$, $A\alpha = \delta^{A^{\#}}$, $B\alpha = \delta^{B^{\#}}$, and

$$A\nu = A^{\#}_{(v:G)}, \quad B\nu = B^{\#}_{(v:G)},$$

respectively.

We have to prove

(i) $\underline{a}\,(A\alpha)\, f_s = (\underline{a}\, f_w)(B\alpha)$ for every $\underline{a} \in \mathrm{dom}\, A\alpha$ and

(ii) $\underline{a}\, f_v \in B\nu$ for every $\underline{a} \in A\nu \subseteq A_v$.

If we substitute in $A^{\#}$ for every $a \in A^{\#}_s$, $s \in S(T)$ the element $a\, f_s \in B_s$ we get a $T^{\#}$-algebra $A^{\#\#}$ and an isomorphism $f^{\#}: A^{\#} \longrightarrow A^{\#\#}$ given by $f^{\#}_s = f_s$ for every $s \in S(T)$ and $f^{\#}_s = Id_{A^{\#}_s}$ else.

Because of $A^{\#\#}{\downarrow}T = B$ condition (2) of Definition 4.2.7 implies

$$B\nu = A^{\#\#}_{(v:G)} \quad \text{and} \quad B\alpha = \delta^{A^{\#\#}},$$

respectively.

Since $f^{\#}: A^{\#} \longrightarrow A^{\#\#}$ is a homomorphism we obtain

$$\underline{a}(A\alpha)\, f_s = (\underline{a}(A\alpha))\, f^{\#}_s = (\underline{a}\,\delta^{A^{\#}})\, f_s = (\underline{a}\, f^{\#}_w)\delta^{A^{\#\#}} =$$

$$= (\underline{a}\, f^{\#}_w)(B\alpha) = (\underline{a}\, f_w)(B\alpha)$$

for every $\underline{a} \in \mathrm{dom}(A\alpha) \subseteq A_w$, and

$$\underline{a}\, f_v = \underline{a}\, f^{\#}_v \in A^{\#\#}_{(v:G)} = B\nu$$

if $\underline{a} \in A\nu = A^{\#}_{(v:G)}$. ∎

4.3 Canon Morphisms

Each canon represents intuitively a certain system of concepts including semantically fixed concepts, represented by sorts and operators specified in enrichments by definitions of the empty canon, parameterized concepts, specified in enrichments by definitions, and open fundamental concepts, specified in enrichments by requirement. One advantage of axiomatic specifications is the possibility of minimizing the conceptual distance between the application area and the formal specification of the intended class of problems. In this way an axiomatic specification makes explicitly visible the conceptual distance between the conceptual level of a certain problem and the conceptual level of a given programming language. In most cases the high complexity of a computer program results from this conceptual distance. Therefore the clarity of a computer program can be improved considerably if the representation of the solving algorithm in terms of the conceptual level of the problem and the reduction of the problem level to the level of the programming language are described independently. These two independent dimensions of the complexity of a computer program are very clearly and very well considered in the system CAT , a system for the structured elaboration of correct programs from structured specifications, in Goguen, Burstall (1980).

A further topic that motivates a mathematically precise description of relations between different conceptual levels, i.e. between canons, is the actualization and the specialization of parameterized specifications. By actualization we understand the substitution of open concepts of a canon by semantically fixed concepts of another canon, and by specialization we understand the substitution of open concepts of a canon by open concepts of another canon with a smaller range of interpretations.

A mathematical representation of the relations between different conceptual levels is the following notion of canon morphisms.

Definition 4.3.1. Let (T_1, Δ_1), (T_2, Δ_2) be canons. A theory-morphism $\varphi : T_1 \longrightarrow T_2$, see Definition 3.5.2, is called a canon morphism from (T_1, Δ_1) to (T_2, Δ_2) denoted by

$$\varphi : (T_1, \Delta_1) \longrightarrow (T_2, \Delta_2)$$

if for every $A \in ALG(T_2, \Delta_2)$, $A\varphi \in ALG(T_1, \Delta_1)$ holds .

A canon (T_1, Δ_1) is called a subcanon of the canon (T_2, Δ_2) or (T_2, Δ_2) an enlargement of (T_1, Δ_1) if T_1 is a subtheory of T_2 and if the inclusion of T_1 in T_2 forms a canon morphism. If (T_1, Δ_1) is a subcanon of (T_2, Δ_2) we write

$$(T_1, \Delta_1) \subseteq (T_2, \Delta_2)$$

for short.

A canon morphism $\varphi : (T_1, \Delta_1) \longrightarrow (T_2, \Delta_2)$ is said to be persistent if

(1) for every $B \in ALG(T_1, \Delta_1)$ there is an algebra $A \in ALG(T_2, \Delta_2)$ freely generated by B according $\varphi : T_1 \longrightarrow T_2$ with $B = A\varphi$,

(2) for every $A \in ALG(T_2, \Delta_2)$, A is freely generated by $A\varphi$ according $\varphi : T_1 \longrightarrow T_2$. ∎

In the case that a canon morphism $\varphi: (T_1, \Delta_1) \longrightarrow (T_2, \Delta_2)$ represents a reduction of a system of concepts described by (T_1, Δ_1) to a system of concepts of a lower level described by (T_2, Δ_2) one can notice that the concept of a canon morphism is not yet adequate. The notion of a reduction includes that a fundamental concept of (T_1, Δ_1) is reduced in general to a composition of concepts that can effectively be built up from the fundamental concepts of (T_2, Δ_2). Unfortunately not every sort name or operator introduced by an enrichment by definition represents a compouned concept that can be built up effectively from the fundamental concepts. The concepts of (T, Δ)-resets and (T, Δ)-algorithms as introduced in Section 4.1 are a first mathematical approximation of computability on canons. But, these versions do not fit together very well with the notion of a canon morphism since canon morphisms map sort names to sort names and not to sets of existence equations. There are several ways of reconciling these both concepts. One way is given by the following definition.

<u>Definition 4.3.2</u>. Let $\varepsilon : (T_o, \Delta_o) \longrightarrow (T, \Delta)$ be a persistent inclusion of a subcanon . A sort name $s \in S(T)$ is called r e c u r s i v e o v e r (T_o, Δ_o) if there is a finite $S(T_o)$-sorted system of variables $v: X \longrightarrow S(T_o)$, a finite set $(v:G)$ of existence equations using operators of $\Omega(T)$, and if there is a term $t \in T(\Sigma(T), v)_s$ such that

(1) (T, G, v) is a representation of a (T_o, Δ_o)-reset,

(2) $\mathrm{dom}\, t^A = A_{(v:G)}$ holds for every $A \in ALG(T, \Delta)$, and

(3) the polynomial function

$t^A : A_{(v:G)} \longrightarrow A_s$

is one-to-one and onto for every $A \in ALG(T, \Delta)$. ∎

Now we can say that a sort name $s \in S(T)$ recursive over a subcanon (T_o, Δ_o) represents a compound concept that can effectively be built up from the concepts given by (T_o, Δ_o).

The following theorem is the mathematical background for the modification of specifications as used in Chapter 1.

<u>Theorem 4.3.3</u>. For every pair $\varepsilon : (T_o, \Delta_o) \longrightarrow (T_1, \Delta_1)$, $\varphi: (T_o, \Delta_o) \longrightarrow (T_2, \Delta_2)$ of canon morphisms there exists a pushout

$$\begin{array}{ccc}(T_o, \Delta_o) & \xrightarrow{\ \varepsilon\ } & (T_1, \Delta_1) \\ \varphi \downarrow & & \downarrow \varphi^{\#} \\ (T_2, \Delta_2) & \xrightarrow{\ \varepsilon^{\#}\ } & (T, \Delta)\end{array}$$

Proof. We recall that a canon morphism $\varphi : (T_o, \Delta_o) \longrightarrow (T_2, \Delta_2)$ consists of a pair of functions

$$\varphi_S : S(T_o) \longrightarrow S(T_2) \ , \quad \varphi_\Omega : \Omega(T_o) \longrightarrow \Omega(T_2) \ .$$

We start with the following constructions

$$\begin{array}{ccc}S(T_o) & \xrightarrow{\ \varepsilon_S\ } & S(T_1) \\ \varphi_S \downarrow & & \downarrow \varphi_S^{\#} \\ S(T_2) & \xrightarrow{\ \varepsilon_S^{\#}\ } & S(T)\end{array} \qquad \begin{array}{ccc}\Omega(T_o) & \xrightarrow{\ \varepsilon_\Omega\ } & \Omega(T_1) \\ \varphi_\Omega \downarrow & & \downarrow \varphi_\Omega^{\#} \\ \Omega(T_2) & \xrightarrow{\ \varepsilon_\Omega^{\#}\ } & \Omega(T)\end{array}$$

with

$$S(T) = (S(T_1) + S(T_2))/\varrho_S$$

$$\Omega(T) = (\Omega(T_1) + \Omega(T_2))/\varrho_\Omega$$

where $S(T_1) + S(T_2)$, $\Omega(T_1) + \Omega(T_2)$ are disjoint unions, ϱ_S is the equivalence relation in $S(T_1) + S(T_2)$ generated by the set of pairs

$$\{(s\varepsilon_S, s\varepsilon_S) \mid s \in S(T_o)\}$$

and ϱ_Ω is the equivalence relation in $\Omega(T_1) + \Omega(T_2)$ generated by the set of pairs

$$\{(\sigma\varepsilon_\Omega, \sigma\varepsilon_\Omega) \mid \sigma \in \Omega(T_o)\} .$$

The functions $\varepsilon_S^{\#}$, $\varphi_S^{\#}$, $\varepsilon_\Omega^{\#}$, $\varphi_\Omega^{\#}$ are defined by

$$s\,\varepsilon_S^{\#} = [s]_{\varrho_S} \quad \text{for every } s \in S(T_2) ,$$

$$s\,\varphi_S^{\#} = [s]_{\varrho_S} \quad \text{for every } s \in S(T_1) ,$$

$$\sigma\,\varepsilon_\Omega^{\#} = [\sigma]_{\varrho_\Omega} \quad \text{for every } \sigma \in \Omega(T_2) ,$$

$$\sigma\,\varphi_\Omega^{\#} = [\sigma]_{\varrho_\Omega} \quad \text{for every } \sigma \in \Omega(T_1) .$$

It is easy to see that

$$\varphi_S \cdot \varepsilon_S^{\#} = \varepsilon_S \cdot \varphi_S^{\#} \quad \text{and} \quad \varphi_\Omega \cdot \varepsilon_\Omega^{\#} = \varepsilon_\Omega \cdot \varphi_\Omega^{\#}$$

Now we define two functions

$$\psi_S : S(T_o) \longrightarrow S(T) \ , \quad \psi_\Omega : \Omega(T_1) \longrightarrow \Omega(T)$$

by

$$\psi_S = \varphi_S \circ \varepsilon_S^{\#} = \varepsilon_S \circ \varphi_S^{\#} .$$

As next we define the arity function and the domain conditions for the operators of $\Omega(T)$.

If $\sigma \in \Omega(T_1)$ and there is an operator $\sigma' \in \Omega(T_o)$ with $\sigma' : w \longrightarrow s$ and $\sigma \varphi_\Omega^{\#} = \sigma' \varepsilon_\Omega \varphi_\Omega^{\#}$ then we set

$$\sigma \varphi_\Omega^{\#} : w \cdot \psi_S \longrightarrow s \psi_S \quad \text{and} \quad def_T(\sigma \varphi_\Omega^{\#}) = (def_{T_o}(\sigma'))\psi .$$

Analogously we define the arity and the domain condition if $\sigma \in \Omega(T_2)$, $\sigma'' \in \Omega(T_o)$ with $\sigma \varepsilon_\Omega^{\#} = \sigma'' \varphi_\Omega \varepsilon_\Omega^{\#}$.

If $\sigma \in \Omega(T_1)$ with $\sigma : w \longrightarrow s$ and there is no $\sigma' \in \Omega(T_o)$ with $\sigma \varphi_\Omega^{\#} = \sigma' \varepsilon_\Omega \varphi_\Omega^{\#}$ then we define

$$\sigma \varphi_\Omega^{\#} = w \cdot \varphi_S^{\#} \longrightarrow s \varphi_S \quad \text{and} \quad def_T(\sigma \varphi_\Omega^{\#}) = (def_{T_1}(\sigma))\varphi^{\#}$$

and analogously we define the arity and domain condition of $\sigma \varepsilon_\Omega^{\#}$ for $\sigma \in \Omega(T_2)$ if there is no $\sigma'' \in \Omega(T_o)$ with $\sigma \varepsilon_\Omega^{\#} = \sigma'' \varphi_\Omega \varepsilon_\Omega^{\#}$.

Straightforward considerations show that the arity function and the domain conditions are defined independently of the choice of representatives.

Since $\Theta(T_o)$, $\Theta(T_1)$, $\Theta(T_2)$ are hierarchical ep-signatures $\Theta(T)$ becomes hierarchical, too.

As next we define the set of axioms $\mathcal{A}(T)$ by

$$\mathcal{A}(T) = \mathcal{A}(T_o)\psi \cup \mathcal{A}(T_1)\varphi^{\#} \cup \mathcal{A}(T_2)\varepsilon^{\#}$$

and so $\varepsilon^{\#} : T_2 \longrightarrow T$, $\varphi^{\#} : T_1 \longrightarrow T$ become evidently theory-morphisms.

Finally we define the set Δ of initial restrictions in T by

$$\Delta = \Delta_1 \varphi^{\#} \cup \Delta_2 \varepsilon^{\#}$$

with

$$\Delta_1 \varphi^{\#} = \{(T', \pi \circ \varphi^{\#} : T'' \longrightarrow T) \mid (T', \pi : T'' \longrightarrow T_1) \in \Delta_1\} ,$$

$$\Delta_2 \varepsilon^{\#} = \{(T', \pi \circ \varepsilon^{\#} : T'' \longrightarrow T) \mid (T', \pi : T'' \longrightarrow T_2) \in \Delta_2\} .$$

It remains the proof of the universal property for the canon morphisms

$$\varphi^{\#} : (T_1, \Delta_1) \longrightarrow (T, \Delta) , \quad \varepsilon^{\#} : (T_2, \Delta_2) \longrightarrow (T, \Delta) .$$

For this purpose let

$$\varphi^{\#\#} : (T_1, \Delta_1) \longrightarrow (T^{\#}, \Delta^{\#}) , \quad \varepsilon^{\#\#} : (T_2, \Delta_2) \longrightarrow (T^{\#}, \Delta^{\#})$$

be canon morphisms with $\varepsilon \circ \varphi^{\#\#} = \varphi \circ \varepsilon^{\#\#}$. Then we obtain functions

$$\delta_S : S(T) \longrightarrow S(T^{\#}) , \quad \delta_\Omega : \Omega(T) \longrightarrow \Omega(T^{\#})$$

uniquely determined by

$$\varepsilon_S^{\#} \circ \delta_S = \varepsilon_S^{\#\#} , \quad \varphi_S^{\#} \circ \delta_S = \varphi_S^{\#\#} .$$

$$\varepsilon^{\#}_{\Omega} \circ \delta_{\Omega} = \varepsilon^{\#\#}_{\Omega} \ , \quad \varphi^{\#}_{\Omega} \circ \delta_{\Omega} = \varphi^{\#\#}_{\Omega}$$

since

$$\varphi^{\#}_{S} : S(T_1) \longrightarrow S(T) \ , \quad \varepsilon^{\#}_{S} : S(T_2) \longrightarrow S(T)$$

$$\varphi^{\#}_{\Omega} : \Omega(T_1) \longrightarrow \Omega(T) \ , \quad \varepsilon^{\#}_{\Omega} : \Omega(T_2) \longrightarrow \Omega(T)$$

satisfy the corresponding universal property with respect to

$$\varepsilon_{S} : S(T_o) \longrightarrow S(T_1) \ , \quad \varphi_{S} : S(T_o) \longrightarrow S(T_2)$$

$$\varepsilon_{\Omega} : \Omega(T_o) \longrightarrow \Omega(T_1) \ , \quad \varphi_{\Omega} : \Omega(T_o) \longrightarrow \Omega(T_2)$$

respectively.

One can see easily that $(\delta_S, \delta_{\Omega})$ forms a canon morphism

$$\delta : (T, \Delta) \longrightarrow (T^{\#}, \Delta^{\#}) \ .$$ ∎

The use of the pushout construction for canon morphisms is based on the following theorem proved by U.L.Hupbach in Hupbach (1980) and (1981).

Theorem 4.3.4. Let the diagram

$$\begin{array}{ccc} (T_o, \Delta_o) & \longrightarrow & (T_1, \Delta_1) \\ \downarrow & & \downarrow \\ (T_2, \Delta_2) & \longrightarrow & (T, \Delta) \end{array}$$

be a pushout of canon morphisms.

(1) An algebra $A \in ALG(T)$ is a (T, Δ)-algebra if and only if $A\varepsilon^{\#} \in ALG(T_2, \Delta_2)$ and $A\varphi^{\#} \in ALG(T_1, \Delta_1)$.

(2) For all algebras $B \in ALG(T_1, \Delta_1)$, $C \in ALG(T_2, \Delta_2)$ with $B\varepsilon = C\varphi$ there is exactly one (T, Δ)-algebra, denoted by $B \oplus C$, with $(B \oplus C)\varphi^{\#} = B$ and $(B \oplus C)\varepsilon^{\#} = C$.

For every pair of homomorphisms

$$f : B_1 \longrightarrow B_2 \ , \quad g : C_1 \longrightarrow C_2$$

with $B_1, B_2 \in ALG(T_1, \Delta_1)$, $C_1, C_2 \in ALG(T_2, \Delta_2)$ and $f\varepsilon = g\varphi$ there is exactly one homomorphism

$$f \oplus g : B_1 \oplus C_1 \longrightarrow B_2 \oplus C_2$$

with

$$(f \oplus g)\varphi^{\#} = f \quad \text{and} \quad (f \oplus g)\varepsilon^{\#} = g \ .$$

(3) If $\varepsilon : (T_o, \Delta_o) \longrightarrow (T_1, \Delta_1)$ is the inclusion of a subcanon then one can assume without loss of generality that $\varepsilon^{\#} : (T_2, \Delta_2) \longrightarrow (T, \Delta)$ is the inclusion of a subcanon, too.

(4) If $\varepsilon : (T_o, \Delta_o) \longrightarrow (T_1, \Delta_1)$ is persistent, then $\varepsilon^{\#} : (T_2, \Delta_2) \longrightarrow (T, \Delta)$ is persistent.

(5) If $\varepsilon : (T_o, \Delta_o) \longrightarrow (T_1, \Delta_1)$ is a persistent inclusion of a subcanon and if every sort name $s \in S(T_1)$ is recursive over (T_o, Δ_o) then every sort name $s \in S(T)$ is recursive over (T_2, Δ_2)

Proof:
To (1): Let be $A \in ALG(T)$ and $(T', \pi \circ \varphi^{\#} : T'' \longrightarrow T) \in \Delta$ with $(T', \pi : T'' \longrightarrow T) \in \Delta_1$. Then $A \models (T', \pi \circ \varphi^{\#} : T'' \longrightarrow T)$ holds if and only if $A(\pi \circ \varphi^{\#}) = (A\varphi^{\#})\pi$ is freely generated by $(A\varphi^{\#})\pi \downarrow T'$, i.e. iff $A\varphi^{\#} \models (T', \pi : T'' \longrightarrow T)$. Analogously, $A \models (T', \pi \circ \varepsilon^{\#} : T'' \longrightarrow T)$ iff $A\varepsilon^{\#} \models (T', \pi : T'' \longrightarrow T)$ for every $(T', \pi : T'' \longrightarrow T) \in \Delta_2$.

To (2): For $B \in ALG(T_1, \Delta_1)$, $C \in ALG(T_2, \Delta_2)$ with $B\varepsilon = C\varphi$ we define $B \oplus C$ as follows:
$(B \oplus C)_s = B_{s'}$ for $s \in S(T)$ if there is an $s' \in S(T_1)$ with $s = s'\varphi_S^{\#}$, and $(B \oplus C)_s = C_{s''}$ for $s \in S(T)$ if there is an $s'' \in S(T_2)$ with $s = s''\varphi_S^{\#}$. Analogously we set $\sigma^{(B \oplus C)} = \sigma_1^B$ if there is a $\sigma_1 \in \Omega(T_1)$ with $\sigma = \sigma_1\varphi_\Omega^{\#}$ and $\sigma^{(B \oplus C)} = \sigma_2^C$ if there is a $\sigma_2 \in \Omega(T_2)$ with $\sigma = \sigma_2\varepsilon_\Omega^{\#}$.
In the same way the homomorphism $f \oplus g : B_1 \oplus C_1 \longrightarrow B_2 \oplus C_2$ can be constructed, i.e. $(f \oplus g)_s = f_{s'}$ if there is an $s' \in S(T_1)$ with $s = s'\varphi_S^{\#}$, and $(f \oplus g)_s = g_{s''}$ if there is an $s'' \in S(T_2)$ with $s = s''\varepsilon_S^{\#}$.

To (3): We have to prove that $\varepsilon_S^{\#} : S(T_2) \longrightarrow S(T)$ and $\varepsilon_\Omega^{\#} : \Omega(T_2) \longrightarrow \Omega(T)$ are injective.
According to the construction of the sets $S(T)$, $\Omega(T)$ and the definitions of the functions $\varepsilon_S^{\#}$, $\varepsilon_\Omega^{\#}$ for $s', s'' \in S(T_2)$

$$s'\varepsilon_S^{\#} = s''\varepsilon_S^{\#}$$

holds if there are $s_1, s_2, \ldots, s_n \in S(T_o)$ with $s' = s_1\varphi_S$, $s_1\varepsilon_S = s_2\varepsilon_S$, $s_2\varphi_S = s_3\varphi_S$, $s_3\varepsilon_S = s_4\varepsilon_S$, ..., $s'' = s_n\varphi_S$.
Since $\varepsilon_S : S(T_o) \longrightarrow S(T_1)$ is injective, we obtain $s_1 = s_2$, $s_3 = s_4$, ... and so

$$s' = s_1\varphi_S = s_2\varphi_S = s_3\varphi_S = \ldots = s_n\varphi_S = s''.$$

In the same way follows the injectivity of $\varepsilon_{\Omega}^{\#} : \Omega(T_2) \longrightarrow \Omega(T)$.

To (4): As a preparation we prove the following statement:
$A \in ALG(T,\Delta)$ is freely generated by $A\varepsilon^{\#}$ according $\varepsilon^{\#}: T_2 \longrightarrow T$ if $A\varphi^{\#}$ is freely generated by $(A\varphi^{\#})\varepsilon$ according $\varepsilon: T_o \longrightarrow T_1$.
For the proof of this assertion let $f: A\varepsilon^{\#} \longrightarrow B\varepsilon^{\#}$ be a homomorphism with $B \in ALG(T)$. Then there is a homomorphism
$f\varphi : (A\varepsilon^{\#})\varphi \longrightarrow (B\varepsilon^{\#})\varphi$, and because of $\varphi \circ \varepsilon^{\#} = \varepsilon \circ \varphi^{\#}$ it holds
$f\varphi : (A\varphi^{\#})\varepsilon \longrightarrow (B\varphi^{\#})\varepsilon$.
Since $A\varphi^{\#}$ is freely generated by $(A\varphi^{\#})\varepsilon$ there is exactly one homomorphism

$$g: A\varphi^{\#} \longrightarrow B\varphi^{\#}.$$

By means of (2) we have a homomorphism

$$f \oplus g : A\varepsilon^{\#} \oplus A\varphi^{\#} \longrightarrow B\varepsilon^{\#} \oplus B\varphi^{\#}$$

and because

$$A = A\varepsilon^{\#} \oplus A\varphi^{\#} \quad \text{and} \quad B = B\varepsilon^{\#} \oplus B\varphi^{\#}$$

there is exactly one homomorphism

$$f \oplus g : A \longrightarrow B$$

with $(f \oplus g)\varepsilon^{\#} = f$. This proves that A is freely generated by $A\varepsilon^{\#}$ according $\varepsilon^{\#} : T_2 \longrightarrow T$.

As next we prove condition (1) of Definition 4.3.1. Let be given $B \in ALG(T_2, \Delta_2)$. Then $B\varphi \in ALG(T_o, \Delta_o)$ and the persistency of $\varepsilon : (T_o, \Delta_o) \longrightarrow (T_1, \Delta_1)$ implies the existence of an algebra $C \in ALG(T_1, \Delta_1)$ such that $C\varepsilon = B\varphi$ and C is freely generated by $C\varepsilon$.
Because of $B \in ALG(T_2, \Delta_2)$, $C \in ALG(T_1, \Delta_1)$ and $B\varphi = C\varepsilon$,

$$B \oplus C \in ALG(T, \Delta)$$

holds according (2).
Since $(B \oplus C)\varepsilon^{\#} = B$ and since $(B \oplus C)\varphi^{\#} = C$ is freely generated by $C\varepsilon = B\varphi$ the statement proved as preparation yields that $B \oplus C$ if freely generated by B and so condition (1) is satisfied.
For the proof of condition (2) of (4.3.1) we assume $A \in ALG(T,\Delta)$. Then $A\varphi^{\#} \in ALG(T_1, \Delta_1)$ and since $\varepsilon : (T_o, \Delta_o) \longrightarrow (T_1, \Delta_1)$ is persistent, $A\varphi^{\#}$ is freely generated by $(A\varphi^{\#})\varepsilon = (A\varepsilon^{\#})\varphi$.
A further application of the statement proved above implies that A is freely generated by $A\varepsilon^{\#}$ according $\varepsilon^{\#} : T_2 \longrightarrow T$ and so condition (2) is also satisfied.

To (5): By means of (3) and (4) we can suppose that

$$\varepsilon^{\#} : (T_2, \Delta_2) \longrightarrow (T, \Delta)$$

is a persistent inclusion of a subcanon.

Let $s \in S(T)$ be any sort name. If there is a sort name $s'' \in S(T_2)$ with $s = s'' \varepsilon_S^{\#}$ then s is evidently recursive over (T_2, Δ_2). Otherwise there is a sort name $s' \in S(T_1)$ with $s = s' \varphi_S^{\#}$.
Since $s' \in S(T_1)$ is recursive over (T_o, Δ_o) there is an $S(T_o)$-sorted system of variables $v: X \longrightarrow S(T_o)$, a finite set of existence equations $(v:G)$ and a term

$$t \in T(\Sigma(T_1), v)_s$$

such that conditions (1),(2),(3) of Definition 4.3.2 are satisfied.
We show that $s \in S(T)$ is recursive over (T_2, Δ_2) with respect to the $S(T_2)$-sorted system of variables $v \circ \varphi : X \longrightarrow S(T_2)$, the finite set of existence equations $(v \circ \varphi : G\varphi^{\#})$ and the term $t\varphi^{\#}$.
For $B \in ALG(T_2, \Delta_2)$ let $A \in ALG(T_1, \Delta_1)$ denote a free T_1-extension of $B\varphi$ which always exists since $\varepsilon : (T_o, \Delta_o) \longrightarrow (T_1, \Delta_1)$ is a persistent inclusion of a subcanon.
According to the proof of (4) $A \oplus B$ is a free T-extension of B and so $(T, G\varphi^{\#}, v \circ \varphi)$ is a representation of a (T_2, Δ_2)-reset.
By the proof of Corollary 3.5.4 we obtain

$$(t\varphi^{\#})^{A \oplus B} = t^{(A \oplus B)\varphi^{\#}} = t^A$$

and

$$(A \oplus B)_{(v \circ \varphi : G\varphi^{\#})} = (A \oplus B)_{(v \circ \varphi^{\#} : G\varphi^{\#})} = ((A \oplus B)\varphi^{\#})_{(v:G)}$$

$$= A_{(v:G)}$$

so that

$$\mathrm{dom}(t\varphi^{\#})^{A \oplus B} = \mathrm{dom}\, t^A = A_{(v:G)} = (A \oplus B)_{(v \circ \varphi : G\varphi^{\#})}$$

Finally

$$(t\varphi^{\#})^{A \oplus B} = t^A$$

is bijective for every $(A \oplus B) \in ALG(T, \Delta)$ since

$$C = C\varphi^{\#} \oplus C\varepsilon^{\#}$$

for every $C \in ALG(T, \Delta)$, and since $(t\varphi^{\#})^C$ is bijective for every $C \in ALG(T, \Delta)$. ∎

Of special interest is assertion (5) of (4.3.4) which roughly says that the relation of effective reducability between concepts, represented by sort names of a canon, is preserved by actualizations and by specializations of open fundamental concepts. This result founds mathematically the approach of stepwise refinement a basic concept in software technology, see Wirth (1971).

REFERENCES

ADJ (1975)
Goguen,J.A., Thatcher, J.W., Wagner,E.G. and Wright,J.B.(1975) Initial algebraic semantics. IBM Research report RC 5364.

ADJ (1978)
Goguen,J.A., Thatcher,J.W. and Wagner,E.G.(1978) An initial algebra approach to the specification, correctness, and implementation of abstract data types. Current Trends in Programming Methodology, Vol.4: Data Structuring (R.T.Yeh,ed.), Prentice-Hall, pp. 80-149.

Benecke,K.(1979) Signaturketten und Operations- und Mengenformen über Signaturketten. Thesis, Technische Hochschule "Otto von Guericke", Magdeburg.

Benecke,K. (1983) Spezifikation parameterisierter Datentypen. Zeitschr. f.math.Logik und Grundlagen d.Math.,29,pp.83-96.

Bergstra,J.A. and Meyer,J. (1983) On specifying sets of integers. to appear in Journal of Information Processing and Cybernetics.

Birkhoff,G. and Lipson,J.D. (1970) Heterogeneous algebras. J. Combinatorial Theory,8,pp. 115-133.

Burmeister,P. (1984) A model theoretic approach to partial algebras, Part I of 'Introduction to theory and application of partial algebras', to appear, Akademie-Verlag, Berlin.

Burstall,R.M. and Goguen,J.A. (1980) The semantics of Clear, a specification language. Proc. of Advanced Course on Abstract Software Specifications,Copenhagen. Springer Lecture Notes in Computer Science, 86, pp. 292-332.

Ehrig,H. (1981) Parameterized specification with requirements. Proc. CAAP'81. Springer Lecture Notes in Computer Science, 112, pp. 1-24.

Goguen,J.A. (1978) Abstract errors for abstract data types. Proc. IFIP Working Conf. on the Formal Description of Programming Concepts, New Brunswick, New Jersey.

Goguen,J.A. and Burstall,R.M. (1980) CAT, a system for the structured elaboration of correct programs from structured specifications. Technical report CSL-118, Computer Science Laboratory, SRI International.

Goguen,J.A. and Burstall,R.M. (1983) Institutions: Abstract model theory for program specification. Draft report, SRI International.

Goguen,J.A. and Meseguer,J. (1981) Completeness of many-sorted equational logic. SIGPLAN Notices 16, 7, pp.24-32.

Grätzer,G. (1979) Universal algebra. 2nd ed. Springer-Verlag .

Hasse,M. and Michler,L. (1966) Theorie der Kategorien. Deutscher Verlag d.Wissenschaften, Berlin.

Hoare,C.A.R. (1972) Proof of correctness of data representations. Acta Informatica 4, pp.271-281.

Höft,H. (1973) Weak and strong equations in partial algebras. Algebra Universalis 3, pp. 203-215.

Hupbach,U.L. (1980) Abstract implementation of abstract data types. Proc. 9th Int. Symp. on Mathematical Foundations of Computer Science, Rydzyna, Poland. Springer Lecture Notes in Computer Science, Vol. 88, pp. 291-304.

Hupbach,U.L. (1981) Abstract Implementation and parameter substitution. Proc. 3rd Hungarian Computer Science Conference, Budapest.

Hupbach,U.L., Kaphengst,H. and Reichel,H. (1980) Initial algebraic specification of data types, parameterized data types, and algorithms. VEB Robotron, Zentrum für Forschung und Technik, Dresden.

Kaphengst,H. (1981) What is computable for abstract data types? Proc. FCT'81, Springer Lecture Notes in Computer Science, Vol. 117, pp. 173-183.

Kaphengst,H. and Reichel,H. (1971) Algebraische Algorithmentheorie. VEB Robotron, Zentrum für Forschung und Technik, Dresden.

Kerkhoff,R. (1970) Gleichungsdefinierbare Klassen partieller Algebren. Math. Annalen 185, pp. 112-133.

Knuth,D.E. (1968) Semantics of context-free languages. Math. Syst. Theory 2, pp. 127-145.

Letitshevski,A.A. (1968) Syntax and semantics of formal languages (Russian). Cybernetics 4,

Lugowski,H. (1976) Grundzüge der Universellen Algebra. Teubner-Texte zur Mathematik, Teubner Verlag, Leipzig.

MacLane,S. (1971) Categories for the Working Mathematician. Springer.

Poythress,V.S. (1979) Partial morphisms on partial algebras. Algebra Universalis 3, pp. 182-202.

Reichel,H. (1969) Kanonische Zerlegung von Funktoren. Math. Nachr. 41, 4-6,pp.361-370.

Reichel,H. (1979) Theorie der Äquoide. Dissertation B, Humboldt University, Berlin.

Reichel,H. (1980) Initially restricting algebraic theories. Proc. 9th Int. Symp. on Mathematical Foundation of Computer Science, Rydzyna, Poland. Springer Lecture Notes in Computer Science, Vol. 88, pp. 504-514.

Rogers,H. (1967) Theory of Recursive Functions and Effective Computability. McGraw-Hill.

Schmidt,J. (1966) A general existence theorem on partial algebras and its special cases. Coll. Math. 14, pp. 73-87.

Shoenfield,J.R. (1967) Mathematical Logic. Addison-Wesley PC.

Slominski,J.(1968) Peano-algebras and quasi-algebras. Dissertationes Mathematicae 62.

Thiele,H. (1966) Wissenschaftstheoretische Untersuchungen in algorithmischen Sprachen I (Theorie der Graphschemata-Kalküle). VEB Verlag der Wissenschaften, Berlin.

Wirth,N. (1971) Program development by stepwise refinement. Comm. of the ACM 14, 4, pp. 221-227.

Zilles,S.N. (1980) Introduction to data algebras. Proc. of Advanced Course on Abstract Software Specifications, Copenhagen.Springer Lecture Notes in Computer Science, Vol. 86, pp.248-272.

Zilles,S.N. (1981) A look at algebraic specifications. Proc. 6th IBM Symposium on Mathematical Foundations of Computer Science.

TABLE OF NOTATIONS

INDEX

ENGLISH TITELS from AKADEMIE-VERLAG BERLIN
in the Series
MATHEMATISCHE LEHRBÜCHER UND MONOGRAPHIEN

NON-LINEAR PARAMETRIC OPTIMIZATION
by B. Bank, J. Guddat, D. Klatte, B. Kummer and K. Tammer
1982. 228 pages - 5 figures - 45,- M
Order number: 762 987 7 (6636)
(Mathematische Lehrbücher und Monographien Abt.II, Monographien, Band 58)

In this book various classes of non-linear parametric optimization problems (including integer problems) are studied theoretically and under the aspect of possible practical applications.
The behavior of the parameter-dependent constraint set, solution set (optimal set) and extreme values of the corresponding optimization problem is examined, this for the main part leads to an analysis of concrete point-set-mappings. Topological and structural própertles are demonstrated for special classes of optimization problems (convex quadratic problems, linear complementary problems with positive semidefinite matrix, non-linear optimization problems with parameter-dependent objective function and mixed-integer problem).
As in linear parametric optimization the concept of local stability sets plays a central role. Special attention is paid to the study of the restriction of the optimal set mapping to a local stability set, thus the behaviour of the optimal set under perturbations of the parameters in a local stability set may be characterized.
The theory developed also provides a foundation for establishing solution procedures involving the simplex technique. On the basis of computing experience only procedures providing solutions in practical cases are selected.

Please order from your bookseller

Akademie-Verlag • Berlin

ENGLISH TITELS from AKADEMIE-VERLAG BERLIN
in the series
MATHEMATISCHE LEHRBÜCHER UND MONOGRAPHIEN

COMPLEX ANALYSIS
Methods, Trends, and Applications

Edited by Eberhard Lanckau and Wolfgang Tutschke
1983. 398 pages - 28 figures - 68,- M
Order number: 763 109 6 (6699)
(Mathematische Lehrbücher und Monographien Abt.II, Monographien, Band 61)

The book written by an international team introduces into the whole sphere of complex analysis. It describes the new methods of complex analysis, which are based on the solution of the inhomogeneous Cauchy-Riemann equations. Those differential equations make it possible to reduce nonholomorphic functions to holomorphic ones.
The book introduces classical and new applications of complex analysis, especially differential equations. The book is written not only for specialists in complex analysis. Moreover it is directed to all mathematicians, physicists and scientists who are interested in analysis, in general.
Authors:
B. Balk, H. Begehr, B. Bojarski, D.L. Colton, A. Džuraev, H. Florian, O. Forster, R.P. Gilbert, K. Habetha, R. Heersink, B. Jesiak, E. Lanckau, J. Lawrynowicz, G.F. Mandžavidze, E. Mues, E.I. Obolašvili, S. Prößdorf, H.-J. Roßberg, G. Tsagas, W. Tutschke, Z.D. Usmanov, V.S. Vinogradov, W.L. Wendland, L. v.Wolfersdorf

Please order from your bookseller

Akademie-Verlag • Berlin

Band 1

Inverse and Improperly Posed Problems in Differential Equations

Proceedings of the Conference on Mathematical and Numerical Methods, held in Halle/Saale (GDR) from May 29 to June 2, 1979

Edited by Gottfried Anger
296 Seiten, 1979

Band 2

Fundamentals of Computation Theory FCT '79

Proceedings of the Conference on Algebraic, Arithmetic and Categorial Methods in Computation Theory, held in Berlin/Wendisch-Rietz (GDR) September 17—21, 1979

Edited by Lothar Budach
576 Seiten, 1979

Band 3

Decidability and Generalized Quantifiers

By Andreas Baudisch, Detlef Seese, Hans-Peter Tuschik, and Martin Weese
XII, 235 Seiten, 1980

Band 4

Zur Elastizitäts- und Thermoelastizitätstheorie

Potentialtheoretische Methoden in Anwendung auf verschiedene Modelle

Von Lothar Jentsch und Johannes Maul
157 Seiten, 1980

Band 5

Systems Analysis and Simulation 1980

Proceedings of the International Symposium held at Berlin (GDR) September 1—5, 1980

Edited by Achim Sydow
496 Seiten, 1980

Band 6

Nichtlineare diskrete Optimierung

Von Eberhard Girlich und M. M. Kowaljow
218 Seiten, 1981

Band 7

Completeness Problems in the Structural Theory of Automata

By Jürgen Dassow
148 Seiten, 1981

Band 8

Darstellung von Lösungen linearer elliptischer Differentialgleichungen

Von Günther Wildenhain
92 Seiten, 1981

Band 9

Introduction to the Spectral Theory of Operators in Spaces with an Indefinite Metric

By I. S. Iohvidov, M. G. Krein, and H. Langer
120 Seiten, 1982

Band 10

Nonlinear Time-discrete Systems

A General Approach by Nonlinear Superposition

By Michael Gössel
115 Seiten, 1982

Band 11

Preprimal Algebras

By Klaus Denecke
162 Seiten, 1982

Band 12

Ergodic Theory and Related Topics

Proceedings of the Conference held in Vitte/Hiddensee (GDR) October 19—23, 1981

Edited by Horst Michel
232 Seiten, 1982

Band 13

Singularly Perturbed Differential Equations

By H. Goering, A. Felgenhauer, G. Lube, H.-G. Roos, and L. Tobiska
176 Seiten, 1983

Band 14

Leben wir in einer Volterra-Welt?

Ein ökologischer Zugang zur angewandten Systemanalyse

Von Manfed Peschel und Werner Mende
233 Seiten, 1983

Band 15

Markov Decision Problems with Countable State Spaces Optimality Criteria — Algorithms — Clustering

By Hans Michael Dietz and Volker Nollau
174 Seiten, 1983

Band 16

Stochastische Verfahren zur Suche nach dem Minimum einer Funktion

Von Ryszard Zieliński und Peter Neumann
133 Seiten, 1983

Band 17

Invertibility and Asymptotics of Toeplitz Matrices

By Albrecht Böttcher and Bernd Silbermann
200 Seiten, 1983

The aim of this book is the initiation of an increasing use of partial algebras especially in computer science. We hope to achiev this by

- presenting examples of increasing complexity of algebraic specifications using essentially partial operations,
- proving that structural induction only yields on partial algebras a complete calculus of abstract programs on abstract data types,
- developing a special kind of partiality, the equational partiality, as well as a sound formal basis for algebraic specifications with partial operations as one proper generalization of universal algebra from total algebras to partial ones.

As distrinct from Part I by P. Burmeister this book does not cover the whole range of generalizing universal algebra to partial algebras.

Es ist die Absicht dieses Buches, partiellen Algebren zu einer wachsenden Verwendung insbesondere in der Informatik zu verhelfen. Dies soll erreicht werden durch

- die Darstellung von Beispielen, mit wachsender Kompliziertheit, von algebraischen Spezifikationen, in denen partielle Operationen wesentlich verwendet werden,
- den Nachweis, daß schon allein die strukturelle Induktion über partiellen Algebren einen vollständigen Kalkül abstrakter Programme über abstrakten Datentypen liefert,
- die Entwicklung einer speziellen Art der Partialität, der Gleichungspartialität, sowohl als eine geeignete formale Basis für algebraische Spezifikationen mit partiellen Algebren, als auch als eine Verallgemeinerung der Universalen Algebra von totalen Algebren für partielle Algebren.

Im Unterschied zum Teil I von P. Burmeister will dieses Buch nicht das ganze Spektrum der Verallgemeinerungen der Universalen Algebra für partielle Algebren überdecken.

ISSN 0138-3019